I0500465

Essays on Safety, Health, and Environment

DEDICATION

I dedicate this book to my loving wife Tammy, and our two sons Fred II and Ted. As proud as I am of my published work that pride pales in comparison to the pride I have in our two sons. They are clearly the best legacy a father could ever ask for.

.

CONTENTS

Introduction

Thank you for choosing this collection of essays. I put them together as a source of my writings on Safety, Health, and Environment (SHE) penned over a period of 20 years. This book does not include writings in which I have the copyright under contract.

My timing for publishing this book is to commemorate the 100th anniversary for what I believe is the seminal event for fire safety. That is the Triangle Waist Factory fire in New York City. This was a waist or shirt factory with mostly young women working for low wages in poor conditions. The fire resulted in 145 workers killed. Many of these young women were immigrants supporting families with their wages.

At the time of this fire, there were labor movements striving to improve the working conditions of these young women. The fire added support to these movements and helped get several fire safety rules passed in New York.

From that event, the safety profession took off. That profession was later combined with occupational health and later still with environmental protection professionals. In some organizations, the word safety leads the title, while in others, environmental lead the title. In some organizations, these duties are done by three people, and in others, all three duties are done by a single person. This reduction to a single person is based on the cost of salaries. Managers believe that the three duties integrated into one person provide cost savings.

I have included essays on training, hazards, the profession, and the professional in this book that I have found applicable during my years in the SHE profession. I think you will find them applicable too. For more information on this and other safety and occupational health topics please visit the bibliography at the end of each essay.

Essay 1-The Occupational Safety and Health Administration

Introduction

President Richard Nixon signed the Occupational Safety and Health Act into law on the 29th day of December 1970 (Della-Giustina, 2000). The act created the three agencies that administer it. They include the Occupational Safety and Health Administration, National Institute for Occupational Safety and Health, and the Occupational Safety and Health Review Commission (Della-Giustina, 2000). The act authorized the Occupational Safety and Health Administration (OSHA) to regulate private employers in the 50 states, the District of Columbia, and territories (Occupational Safety and Health Act, 1970).

OSHA was established in 1971 and is headquartered in Washington, District of Columbia with regional offices spread across the nation (Fanning, 2003). OSHA develops safety standards in the Code of Federal Regulation and enforces those safety standards through compliance inspections conducted by Compliance Officers. Furthermore, this agency provides training and publications at little or no cost (Fanning, 2003).

Purpose

OSHA is an agency under the Department of Labor whose purpose is to ensure safe and healthful workplaces in America. It fulfills its mission through the implementation of a five-year plan (Department of Labor, 2003). The agency's vision is that "Every employer and employee in the nation recognizes that safety and health adds value to American businesses and workplaces and workers lives" (Department of Labor, 2003). The original plan was for OSHA to oversee 50 state plans, with OSHA funding 50% of each plan. Unfortunately, it has not worked out that way.

There are currently 26 approved state plans but no other states want to participate. OSHA manages the plan in the states not participating.

OSHA has three goals: (1) reduce occupational hazards through direct intervention, (2) promote a safety and health culture through compliance assistance, cooperative programs, and strong leadership, and (3) maximize OSHA's effectiveness and efficiency by strengthening its capabilities and infrastructure (Department of Labor, 2003).

Structure

OSHA operates through ten regional offices that are further broken down into districts. They are organized into three sections: compliance, training, and assistance. OSHA's primary efforts are built on what they call "a strong, fair, and effective enforcement program" (Department of Labor, 2003). OSHA aids employers who want to do the right thing and focuses enforcement resources on high-hazard industries. Each state has consultation services available for small businesses that are funded by OSHA. In addition, OSHA has 73 specialists in local offices to provide tailored information and training to employers and employees (Department of Labor, 2003).

OSHA also has an Alliance Program that enables trade or professional organizations, businesses, labor organizations, educational institutions, and government agencies to collaborate with OSHA. Representatives from OSHA and the organization sign a formal agreement with goals that address training and education, outreach and communication, and promoting the national dialogue on workplace safety and health. Furthermore, OSHA has a Strategic Partnership Program that targets construction, shipbuilding, food processing, logging, silica mining, and nursing homes. This program zeros in on specific hazards or specific geographic areas. OSHA also has a Voluntary Protection Program (VPP). Potential VPP worksites apply to OSHA. A review of the application is conducted and, if approved, OSHA conducts an on-site inspection. If the organization passes the inspection, it is given VPP status. This means OSHA will not conduct annual inspections and normally not visit a site unless there is a fatal accident or an employee complaint. What does the organization get for all this work? VPP sites have injury and illness rates more than 50% below the average for their industry. This relates to savings millions of dollars each year.

OSHA manages a Susan B. Harwood grant that provides money to nonprofit companies to provide training. The Occupational Safety and Health Act of 1970 along with the Department of Labor, Health and Human Services, Education, and Related Agencies Appropriation Act authorize this grant. The grant provides funds to train workers and employers to recognize, avoid, and prevent safety and health hazards in the workplace (Federal Register, 2002). In 2002, there were Targeted Topic Grants and Institutional Competency Building Grants. These grants were awarded for 12 months and averaged $200,000. In 2003, there were Targeted Topic Grants, OSHA Training Material Grants, and Ergonomic Guidelines Training Grants. These grants were awarded for 12 months and averaged $150,000 each (Federal Register, 2003).

Budget

OSHA began fiscal year 2003 (the Federal fiscal year (FY) runs from October 1 - September 30) with a staff of 2,303 including 1,123 inspectors. The agency's budget request was for $454 million. In his press release, OSHA Director John Henshaw stated, "This budget will give us the resources we need to help ensure workers' safety and health, while maintaining fiscal responsibility." He went on to say, "It is a strong and sound budget that supports our priorities of leadership, effective enforcement, outreach and education, and partnerships" (OSHA, 2002).

In the budget, OSHA received $448.7 million. This budget included $60.3 million for expanded outreach activities and compliance assistance, which encompassed training and information exchanges and technical assistance to employers. This also included an increase of $250,000 for new computer-based outreach products, $500,000 to increase technology-based training, and $500,000 to improve compliance assistance training for OSHA's staff.

The enforcement budget provided $161.1 million to focus resources on activities that had the greatest impact. OSHA targeted inspections on the worst hazards at the most dangerous workplaces. There should have been 1,300 more inspections in FY 2003 than in previous years.

The new budget kept state program activities at FY 2002 levels. In addition, the budget included an increase of $1.5 million to state consultation programs for small businesses. Finally, $14 million was allocated for the development, review and evaluation of standards. Figure 1 shows a side-by-side comparison of the FY 2002 and FY 2003 budgets.

Budget Items (in millions of dollars)	FY 2002	FY 2003	Change
Safety and Health Standards	$15.5	$14.2	$-1.3
Federal Enforcement	161.8	161.1	-0.7
State Programs	89.8	89.8	0.0
Technical Support	19.6	20.2	0.6
Federal Compliance Assistance	58.8	60.3	1.5
State Consultation Grants	51.0	52.5	1.5
Training Grants	11.2	4.0	-7.2
Safety and Health Statistics	26.2	25.7	-0.5
Executive Direction and Administration	9.0	9.2	0.2
Pension and Health Cost	13.7	11.7	-2.0
Total Budget Authority	$456.6	$448.7	$-7.9

Figure 1 – OSHA Budget Comparison (U.S. Department of Labor, 2003).

Strengths

OSHA provided funding for a lot of personnel from the workforce to be trained. In FY 2002, OSHA trained 3,085 students at its training institute in Des Plaines, Illinois. In addition to this training facility, OSHA had 12 Education Centers around the United States where 14,500 students were trained. OSHA also provided grants to colleges, universities and non-profit organizations, which trained another 65,000 students. OSHA also provided approval for outreach centers to conduct training, accounting for 254,403 students trained (Department of Labor, 2003).

OSHA conducted 17,726 consultations in FY 2002. Eight national alliances were completed as well as 167 Strategic Partnerships. These programs affected 3,550 employers and 200,000 employees. The Voluntary Protection Programs had 637 sites with more than 180 industries with 416,000 employees (Department of Labor, 2003).

In FY 2002, OSHA conducted 20,511 inspections of high-hazard facilities. This was 55% of OSHA's inspections based on the hazard potential of the facility with 57% of its inspections on Construction Companies. This was a high hazard area and deserving of the attention. Another strength was that OSHA funded 50% of approved state programs. The programs at the state level conducted another 58,402 inspections in FY 2002. OSHA issued citations for $72,827,278 in FY 2002 (U.S. Department of Labor, 2003).

OSHA promulgates standards within the Code of Federal Regulation as 29, Labor. These standards are intended to clarify the standard for maintaining a safe and healthful workplace. These standards are also used to train personnel.

Weaknesses

In FY 2002, OSHA Compliance Officials conducted 37,493 Inspections. OSHA conducted 9,007 inspections based on complaints or accidents in the workplace. This was a weakness because 24% of OSHA's inspections were conducted after an accident occurred or an employee failed to get the hazard resolved at the organizational level. OSHA-approved state programs conducted another 58,402 inspections in FY 2002. OSHA issued citations for $72,827,278 in FY 2002. This was money taken from industries and given to the government. Industry cannot afford to lose that much money (Department of Labor, 2003).

William Martin and James Walters explained in their book (2001) that "More than 56% of the U.S. workforce in private industry is employed in business establishments with fewer than 100 employees. Prevention of occupational injury and illness is often difficult in these establishments because small businesses generally have few safety and health resources, cannot usually hire staff devoted to safety and health activities, and often lack the ability to identify occupational hazards and conduct surveillance. OSHA recognizes these special challenges to safety and health in small business establishments." Despite OSHA recognizing these challenges, little was done and they were often a drain on the financial resources of small businesses.

There is a great deal of confusion between the rules and standards that OSHA publishes, the ones states develop, and consensus standards that are promulgated. Neville Tompkins points this out very clearly in his book (2001), saying, "A multitude of regulations have been enacted at the federal, state, county, and municipal levels with the intention of possibly affecting the safety and health of men and women in U.S. workplaces." With all these rules, it is nearly impossible for an employer to know all the regulations governing the workplace.

6

External Influences

OSHA's focus should be to allow for commerce while protecting employees. However, this work is difficult because many people work to undermine this focus. Large organizations and societies lobby Congress to get rules changed to benefit them. This often results in watering down regulations and standards. Furthermore, many in Congress want OSHA to function as a compliance organization while others expect them to be preventive. This dichotomy takes away some effectiveness in that the two are not mutually compatible. There are also employee unions and societies that work to protect employees that want OSHA to be harder on employers and force more regulations and standards to protect the employees. These societies want just the employer looked at while disregarding any role for the employee.

OSHA is often viewed in the press as ineffective and often downright incompetent. In addition, they have been blamed, in the past, as the reason why some manufacturers have moved their factories out of the United States to countries with less regulation. Unfortunately, there is never an adequate debate on these issues on the news or any other open forum.

Impact

OSHA is the federal agency charged with reducing work-related injuries and illnesses, and to its credit, workplace fatalities have been reduced by 50% and injuries and illnesses have been reduced by 40% since OSHA was created. This occurred while the workforce doubled. In FY 2001, the national occupational injury and illness rate dropped to 5.7 cases per 100 workers. This was the lowest level since the U.S. began collecting this information and was part of an eight-year downward trend (Department of Labor, 2003). There were 5,900 worker deaths in 2001, less than one percent fewer than in 2000.

Conclusion

OSHA, like any federal agency, gets its mandate from Congress and is funded with directions for budget implementation. It has at times been mismanaged and ineffective. The director in 2003 was a competent safety professional who seemed to be focused on changing the image of the agency to one of a partner in preventing accidents rather than one who just policed the workplaces of America. OSHA does a lot of good work in training and assisting employers in preventing accidents while at the same time citing and punishing those that don't. OSHA has weaknesses but those are overshadowed by their successes.

Bibliography

Della-Giustina, Daniel E. (2000). Developing a Safety and Health Program, New York: Lewis Publishers.

Fanning, Fred E. (2003). Basic Safety Administration: A Handbook for the New Safety Specialist, Chicago: American Society of Safety Engineers.

Federal Register No. 68:32773-32779 (2003). Susan Harwood Training Grant Program, FY 2003 Budget. Retrieved December 3, 2003 from http://www.osha.gov/pls/oshaweb/owadisp.show_document?p_table=FEDERAL_REGISTER&p_id=17848.

OSHA News Release (2002). President's fiscal year 2003 Budget Request for OSHA Emphasizes Continued Outreach, Enforcement. Retrieved on December 3, 2003 from http://www.osha.gov/pls/oshaweb/owadisp.show_document?p_table=NEWS_RELEASES&p_id=1182&p_text_version=TRUE.

Tompkins, Neville C. (2001). Basics of Safety and Health, Chicago: National Safety Council Press.

U. S. Department of Labor, Occupational Safety and Health Administration. Retrieved on November 28, 2003 from http://www.osha.gov.

References

Federal Register No. 67:6024-36028 (2002). Susan Harwood Training Grant Program, FY 2003 Budget. Retrieved December 3, 2003 from http://www.osha.gov/pls/oshaweb/owadisp.show_document?p_table=FEDERAL_REGISTER&p_id=17290.

Martin, William F. and James B. Walters (2001). Safety and Health Essentials for Small Businesses, Boston: Butterworth and Heinemann.

Occupational Safety and Health Act (1970). United States Congress, Washington District of Columbia: U.S. Government Printing Office.

Essay 2-The Great "Ouchdoors"

Introduction

Public employees work in a variety of occupations. Many of those occupations require them to work out-of-doors amid the various forms of wildlife in this great nation. Some people enjoy this work because of the outdoor wonders. Unfortunately, many of these wonders bite, infect, scratch, or worse. There are a variety of snakes, poisonous plants, spiders, ticks, and mosquitoes that call the outdoors home. The key to avoiding problems is to know what animals, insects, and plants to watch out for. This essay will provide the reader with basic information to help spot these troublesome "critters" and prevent them from harming public employees working in the great "Ouchdoors."

Mammals

The bobcat is a local animal to many areas although it exists in small numbers. The bobcat often looks like a domestic cat; however, it is quite fierce and is equipped to kill animals as large as deer. The bobcat normally eats rabbits, ground squirrels, mice, and wood rats. The Bobcat roams freely at night and is often out during the day. It lives in crevices, caves, dense thickets of brush, or hollowed logs or trees.

Employees may also come across a coyote. The coyote looks like an unkempt family mutt; however, it is not. The coyote weighs between 15 and 45 pounds and is normally 40-60 inches long from snout to tail. They can be 15-20 inches tall at the shoulder. Female coyotes often give birth to six pups and a coyote can live for 15 years in the wild. They typically eat small mammals, insects, reptiles, fruit, and carrion. Despite the long-life span, nearly one out of five pups does not survive their first year. The coyote can run almost 40 mph and is normally afraid of humans, but when the female coyote is protecting her young, she may be aggressive.

Reptiles

Snakes are another of nature's wonders, unless they bite. The average snake does not set out to bite a person. We are much too big; however, if provoked, the snake will bite humans, even very big humans.

The Copperhead is a very common snake in many areas. Like all snakes they like to hide during daylight and are very active at night. You can also see snakes during the early morning hours sunning themselves. They are cold-blooded creatures and can use the sunlight to warm themselves. The Copperhead is normally found in and around wooded areas with a lot of insects, small mammals, and water to drink.

Another snake found in many areas is the Cottonmouth. This snake is much darker in color and prefers to live around water. A local pond, creek, or river suits them just fine. When they open their mouth the inside is a bright white. You never want to see them open their mouth. It usually means they have been provoked. They are like most other snakes and live off insects and perhaps some small mammals. This snake is at home in the water and is an excellent swimmer. They can also jump from the water into a small boat if provoked. Don't ask me how I know this, just trust me.

The Mississauga Rattlesnake is like the other snakes addressed in this essay. It is a darker rattlesnake and blends in very well in shadowy wooded areas. This snake feels right at home under a stray piece of wood or another object under which it can hide. It makes a pronounced rattling noise from its tail when agitated.

Next is the Timber Rattlesnake. This snake is a little brighter in color and still blends in well with wooded areas. This snake will also make a rattling noise with its tail when it is agitated. These and other snakes will normally lash back when picked up and try to bite anything or anyone near them.

The Western Pygmy Rattlesnake is smaller than the other rattlesnakes. It also has a little different color scheme. All snakes should be left alone. The snakes highlighted here are the most dangerous and should be avoided altogether.

Plants

There are also several plants that employees need to know about. They will want to identify them by sight and avoid them.

The first plant is poison ivy. This is a bright green plant with three leaves. It grows anywhere it can find sunlight and a little water. Contact with this plant can cause a severe rash. Some people do not have to encounter the plant to get the rash. Some people can simply encounter other objects that have touched the plant and experience the rash.

Then there is poison oak. This is another plant that can cause a rash. It works like poison ivy. It should also be avoided whenever possible.

Poison sumac is the third plant in many areas that can cause a rash. These plants can be found and are easy to get to. Some yards may even have them.

Insects

There are also some insects in most areas to be aware of. First is the spider. There are many kinds of spiders and they can be found inside and outside. There are a few that are more dangerous than others.
The Brown Recluse is maybe the most dangerous. Its bite causes the skin around the bite to die and venom takes several weeks to work its way through the cycle. This spider likes to be around water sources and can often be found in porta-potties (modern outhouses). It can be identified by its brown color.

Next is the Black Widow. This is another spider that has a dangerous bite. Employees can recognize it by the distinct black color with the red hourglass on its back. It is found in many of the same areas that all spiders are found like water, darkness, out of the way places.

Then there is the multi-talented tick. The vampire of the insect world is known to most of us as a freeloader on pet dogs and cats; however, it can and does attach itself to humans. There are several different kinds of ticks. The larger ticks are the normal, while the smaller full-grown ticks are usually the deer ticks. There are a lot of ticks in the woods in most areas and something as simple as a picnic in your own backyard can bring people in contact with them.

There are fleas in most areas. These are normally seen on pet dogs and cats; however, they also seem to like carpet too. If they can get into a house or office in large numbers they can bite humans and cause small welts around the ankle area.

Let's not forget the ever-popular mosquito. Employees will find more mosquitoes than they ever wanted. Mosquitoes like water and can be found in and around water sources. They prefer stagnant water and will make a home in an old flower pot, tire, hole, or bucket. Once they find a home they breed to their hearts' content. Using repellents and eliminating sources of stagnant water can help keep the population down.

Summary

Public employees work in a variety of occupations. Many of those occupations require them to work out-of-doors amid the various forms of wildlife in this great nation. Unfortunately, many of these wonders bite, infect, scratch, or worse. There are a variety of snakes, poisonous plants, spiders, ticks, and mosquitoes that call the outdoors home that should be avoided always. This essay provided the reader with information to spot these troublesome "critters" and prevent them from harming public employees working in the great "Ouchdoors."

Note: This article was originally published in the Perspectives Newsletter, Volume 9, Number 3, 2010 by the American Society of Safety Engineers, Council on Practices and Standards.

Essay 3-Inclement Weather Road Condition Guidelines

Introduction

If this year is anything like years past, there will be more opportunities for public employees to be plowing snow, clearing sidewalks, and perhaps more importantly, providing fire and ambulance services to their communities in inclement weather. The public expects the government to provide essential services even under extreme weather conditions. This can put a strain on public employees and contractors hired to do the work. There are even many locations that depend on volunteer fire and ambulance services. The hazards of driving in extreme weather can be identified and intelligent decisions made as to who will drive under what conditions. This essay will highlight a rating scheme that includes a risk management matrix to help protect public employees driving in winter weather.

Rating Driving Conditions

Road conditions can be rated on a continuum from green or normal driving conditions to black or dangerous. For discussion purposes, it is best to start with a description of the types of road conditions. GREEN road conditions include dry road surfaces, with no ice or snow, visibility greater than 150 feet and a temperature greater than 35° Fahrenheit (F). Most people would describe these driving conditions as normal. These conditions normally won't cause or contribute to a motor vehicle accident. These conditions normally call for unrestricted vehicle dispatches; however, drivers should be reminded to observe normal precautions and speed limits. Decisions to dispatch vehicles should come from a first line supervisor.

As weather conditions start to impact driving, they are usually identified as AMBER. At this point normal road conditions, temperature and visibility do not exist. The road surface is wet from rain; or the road surface contains packed snow or slush, usually less than four inches; or the road surface has patches of ice or black ice; or visibility is reduced to

13

between 60 and 150 feet; and the temperature is between 30-35° F. At this point drivers must:

- Increase driving times
- Drive 10 miles per hour (MPH) below the posted speed limit to maintain traction
- Increase following distances to allow for safe stopping

Driver experience should be considered when determining whether to dispatch vehicles under Amber conditions. The decision to dispatch should be made by the second level supervisor.

As the weather creates worse road conditions, the description of RED is more appropriate. At this point, water is flooding or snow is drifting across the road surface, or there is sheet ice, or the snow depth is greater than four inches. Visibility is less than 60 feet and the temperature is below 30° F. Only mission-essential and emergency-essential vehicle dispatches should be authorized. Driving above 10-15 MPH could cause vehicles to lose traction and safe stopping distances are significantly increased. A risk assessment should be completed before dispatch to determine if the risks of the vehicle trip are worth taking. The decision to dispatch should now be made by a third level supervisor.

If weather conditions continue to get worse and impact the road surfaces, the term BLACK is applied. The road surfaces are now heavily flooded or heavy drifting of snow is occurring. There are extremely large areas of sheet ice and the snow might be greater than six inches deep. The visibility is now down to 45 feet and the temperature is below 10° F. At this point, the dispatch of vehicles is limited to emergency-essential vehicles (police; fire; ambulance; snow and ice removal (SNAIR) vehicles, and emergency engineering services). Chiefs of appropriate offices (police, fire, medical activity, and public works) should authorize dispatch of vehicles. A risk assessment should be completed before dispatch to identify control measures to reduce risks.

Road Status Condition	Risk (Authority to Proceed)	WEATHER CRITERIA					
		Road Surface	Snow	Ice	Snow Depth	Visibility	Temperature
Green	Normal Operations (1st Line Supervisor)	Dry	None Blowing Powder	None	None	>150'	>35°
Amber	Cautious Operations (2nd Line Supervisory)	Wet	Packed Slush	Patches Black Ice Slush	<4"	60-150'	30°-35°
Red	Restricted Operations (3rd Line Supervisor)	Flooded	Drifting	Sheet Ice	>4"	<60'	<30°
Black	Prohibitive Operations (Chief of Office)	Heavily Flooded	Heavy Drifting	Extreme Sheet Ice	>6"	<45'	<10°

Figure 1 – Adapted from the U.S. Army Road Conditions Risk Criteria

Figure 1 is a matrix adapted from one used in Europe by the US Armed Forces. The matrix is a short and easy-to-use aide to determine who should be driving vehicles and when. The author has a great deal of experience with this matrix and believes it is the best one published.

Summary

As winters go, weather seems to be getting more severe. Now is the time to prepare for this winter's weather. One area of preparation is identifying who will be allowed to drive vehicles when and under what conditions. The more severe the weather, the fewer people should be able to drive vehicles. There will always be exceptions for SNAIR, police, fire, and medical services. The public expects the government to provide essential services even under extreme winter weather conditions and it is up to public employees and contractors to get the job done despite the strain it places on them. As we look at the road hazards of winter, we should not forget about the volunteer fire and ambulance services. This essay has identified the hazards of driving under extreme winter weather. Recommendations were also included to help supervisors make informed and intelligent decisions about who will drive under what conditions. This essay also highlighted a rating scheme that included a risk management matrix to help protect employees.

References

Change 2 to US Army Europe Regulation 385-55, November 11, 1995.

US Army Europe (Forward) Policy Memorandum, "Vehicle Operations under Adverse Weather Conditions", January 6, 1996.

Note: This article was originally published in the Perspectives Newsletter, Volume 10, Number 1, 2010 by the American Society of Safety Engineers, Council on Practices and Standards.

Essay 4-Preventing Cold Weather Injuries

Introduction

During the winter months in North America, winter hazards abound. Public employees usually work during these periods, keeping the roads open, fixing frozen and broken pipes, and completing a variety of other tasks. This exposure puts them at risk of cold weather injuries. Individuals must know the warning signs of cold weather injuries and heed them. The author suggests the best way to prevent or reduce the occurrence of cold weather injuries in public employees is to tell them what cold weather injuries are, how they occur, how serious they are, first-aid treatment, and what they can do to prevent them. This essay will highlight just that information and help public employees prepare for cold weather.

Wind Chill

Before discussing the prevention of cold weather injuries, it is important to understand the concept of wind chill. This is important because the human body is continually producing and losing heat. When the wind blows across the body, it removes heat, making the body susceptible to cold weather injuries. The combined effect of wind and temperature is expressed as an equivalent temperature; see Figure 1 on the next page. Locate the ambient temperature across the top of the figure and run a finger down to the wind speed, and that is the temperature that the body is feeling. For example, if it is 15° Fahrenheit (F) and the wind is blowing 30 miles per hour, the body is experiencing a temperature of -5° F, which is much more dangerous to the body than 15° F.

	Temperature							
W i n d S p e e d	Calm	40	35	30	25	20	15	10	5
5	36	31	25	19	13	7	1	-5
10	34	27	21	15	9	3	-4	-10
15	32	25	19	13	6	0	-7	-13
20	30	24	17	11	4	-2	-9	-15
25	29	23	16	9	3	-4	-11	-17
30	28	22	15	8	1	-5	-12	-19

Figure 1 – Adapted from U.S. Army Wind Chill Chart

Cold Weather Injuries

Cold weather injuries come in various shapes and sizes, so to speak. Hypothermia, Frostbite, and Trench Foot are the most common. There are often extenuating factors that increase one's risk of experiencing a cold weather injury or perhaps making an injury worse, and these include:

- Acclimation to cold weather
- Length of exposure
- Previous cold injuries
- Use of prescription drugs
- Consumption of alcoholic beverages

The next few paragraphs will highlight the different types of cold weather injuries and provide useful information for public employees. It is important to remember that it is always easier to prevent a cold weather injury than to treat one after it occurs.

Hypothermia

Hypothermia is a lowering of the body temperature caused by exposure to the cold. It is aggravated by moisture and wind. Hypothermia occurs when the body is unable to produce heat as quickly as it is being lost. Most hypothermia cases develop in air temperatures between 30 and 50° F. It is important to note that a person will die if the internal body temperature drops below 78.6° F, making this a deadly injury. The moment one begins to lose heat faster than the body can replace it, they experience exposure. This is followed by the body taking drastic measures to conserve its energy and maintain the temperature of internal organs. As the body's core temperature drops, the symptoms shown in Table 1 develop.

When the Body Temperature Is	The Body Reaction Is
96-98° Fahrenheit	Shivering becomes intense and uncontrollable. The ability to perform complex tasks is impaired.
95-91° Fahrenheit	Violent shivering persists. Difficulty in speaking, sluggish thinking, and amnesia start to appear.
85-81° Fahrenheit	Victim becomes irrational, loses contact with the environment, and drifts into a state of suspended or deadened sensibility. Muscular rigidity occurs. Pulse and breathing slow. Victim becomes unconscious. Victim does not respond well to the spoken word. Most reflexes stop functioning. Heartbeat becomes erratic.
Below 78° Fahrenheit	Cardiac and respiratory failure occur and death follows

Table 1 – Human Body Reactions to Lowered Body Temperature

The treatment of hypothermia consists of stopping or at least reducing the heat loss from the victim's body. This is followed by adding heat to the victim's body. This is best done by:

- Getting the victim to a sheltered area even if it is a warm vehicle.

- If the victim is wet, replacing the victim's wet clothes with dry ones.
- Giving the victim warm, non-alcoholic drinks.
- Getting the victim to a hospital or medical clinic as soon as possible.

The prevention of hypothermia consists of:

- Dressing properly in layers (wear a hat and gloves or mittens).
- In rain, choosing rain gear that works against wind-driven rain.
- Using the buddy system or working in pairs so help is nearby if needed.
- Knowing the weather and taking precautions based on the forecast.

Frostbite

Frostbite is the freezing of the skin and tissue of a body part exposed to temperatures of 32° F or below. The first symptom is an uncomfortable aching sensation, tingling, or stinging. If the condition can continue, numbness sets in. The skin initially turns red, later becoming pale gray or waxy white. In extreme cases, frostbite can be very serious and result in the loss of or permanent damage to a body part. People have lost fingers and toes from frostbite.

Frostbite occurs superficially and deep. Treatment depends on the degree of frostbite injury. The longer a body part has been without feeling, the more severe the frostbite. If the time is very short, the frostbite is probably superficial. Otherwise, you should *assume* the injury is deep and therefore, serious. In cases of deep frostbite, seek emergency medical treatment immediately. While waiting for emergency medical care, protect the frozen part of the body from further injury by heeding these do's and don'ts:

- Do keep the unfrozen parts of the body warm.
- Don't thaw frozen body parts by rubbing, bending, or massaging.
- Don't soak the frozen area in either cold or warm water.
- Don't rub the frozen body part with snow.
- Don't expose the frozen body part to hot air, engine exhaust, or

open fires.

- Don't use ointments or salves.

The prevention of frostbite consists of:

- Dressing properly in layers to keep the body warm.
- Always wearing a hat and gloves or mittens.
- Avoid wearing clothing that interferes with circulation. Tight-fitting shoes, socks, and gloves are especially dangerous.
- In rain, choosing rain gear that works against wind-driven rain.
- Using the buddy system or working in pairs so there is help nearby if needed.
- Knowing the weather and taking precautions based on the forecast.
- Exercising your face, fingers, and toes to keep them warm.

Trench Foot

Trench foot is caused by prolonged standing in water or by having wet feet for hours while exposed to a temperature just above freezing. The stages of trench foot are:

- Early stages - feet and toes are pale, numb, and stiff, while walking becomes difficult.
- Later stages - feet and toes become red, swollen, and warm, which can include flesh dying.

There are several reasons the feet are susceptible to trench foot that include:

- The feet are far from the heart, causing the heart to pump blood a long distance to warm the feet.
- When standing for long periods, blood circulation can slow.
- It is easy for the feet, even in waterproof boots, to get wet.
- Tight socks or tight-fitting boots can restrict circulation.

The best prevention for Trench Foot is a good pair of shoes or boots and socks that fit properly. Here are a few dos and don'ts:

- Do dry wet feet as soon as possible and put on dry socks.

- Do dry the inside of boots and shoes.
- Do seek medical attention as soon as foot problems occur.
- Do exercise the feet by stamping, stepping back and forth while flexing and wiggling the toes when working out in the cold.
- Do handle feet gently.
- Do wash feet carefully using mild soap and water.
- Do dry and elevate the feet, leaving them uncovered and at room temperature.
- Don't rub or massage them.
- Don't restrict blood circulation by wearing tight socks or lacing shoes too tightly.

Summary

During the winter months in North America, winter hazards abound. With public employees, out and working in the hazardous winter weather, it is important to tell them what cold weather injuries are, how they occur, how serious they are, first-aid treatment, and what can be done to prevent them. One last point the author would like to highlight is a mnemonic used by the U.S. Army. COLD is a good memory device for the use and care of cold weather clothing and footwear that can be used by employees as the first line of defense against exposure to cold weather. The mnemonic stands for:

Keep it **C**lean
Avoid **O**verheating
Wear it **L**oose in Layers
Keep it **D**ry

Note: This article was originally published in the Perspectives Newsletter, Volume 10, Number 1, 2010 by the American Society of Safety Engineers, Council on Practices and Standards.

Essay 5-Severe Weather Season

Introduction

Each year we are faced with thunderstorms, lightning, hail, and tornadoes. Each storm is unique; however, there are things that you can do to prepare yourself. Get to know what can happen where you live. For example, thunderstorms with lightning, hail, and high winds occur frequently through the spring and summer months in the Midwest. Tornadoes, which are more dangerous, do not occur as often.

Thunderstorms

Thunderstorms are dangerous and can result in property damage and personal injuries. To prepare for thunderstorms, make a family severe weather plan. Plan how each family member will contact the others if a storm occurs when you and your family members are at work and school. It is also good to have a common meeting place just in case your home is damaged so badly you can't go back to it. Build a home storm kit that includes a flashlight (extra batteries), a small battery powered radio (extra battery), a blanket, water (1-2 gallons), a first-aid kit, candles, matches, and spare medications that family members might need to take.

Identify a place in your home that will be safe. This could be a basement, room inside the center of the home, or a bathroom. Monitor the path and severity of storm warnings on television. Be prepared to switch to your radio if you lose television reception or power goes out. When weather approaches, you should plan for the lights to go out. Know where the candles are and how to light them quickly. If you are using a well, plan to lose water. Be prepared to ration use of the toilet and fill the bathtub with water that can be used to flush the toilet. Also, get out the water you have stored for drinking.

You are normally safer inside of a structure in a thunderstorm than outside. However, if you are in a motor vehicle, it may be your only means of protection. It is safer to stay in the vehicle if the storm does not include a tornado. You should have a severe weather kit for your car that includes a flashlight (extra batteries), a blanket, water (1-2 pints), and a first-aid kit. After the storm, remain with your car if it will not start. You will normally get help faster by staying with your vehicle.

If you see lightning or hear thunder, seek protection. If you are in a building, stay off the land line phones and don't touch any water pipes. Both can conduct electricity if they are struck by lightning. Stay in your vehicle and don't touch any metal parts, which also conduct electricity. If you are outside, seek shelter in a building, motor vehicle, under a park cover, or a group of trees using the lightning strike position. Don't seek shelter under a single tree. That single tree is more likely to be struck by lightning. Don't lie or sit on the ground. Lightning often runs along the ground. The more body contact you have with the ground the more likely it is you will be shocked. The lightning strike position is to squat down on both feet, wrap your arms around your knees and rest your head on your knees.

Tornadoes

Seek immediate shelter if a tornado is coming and always take tornado watches and warnings seriously. Know what county you live in and the counties that border yours. Note the direction that the storm is moving as you watch television or listen to radio. If you hear a siren, seek immediate shelter and use your portable radio to learn about the danger. Don't come out of the shelter before the storm is over. Teach your children to return home immediately or seek shelter in a neighbor's home. This will prevent a parent from going out in the storm to look for a child. It is also dangerous to go out in a storm to rescue pets.

Summary

Severe weather cannot be prevented, but you can reduce the impact that it has on you and your family. By taking steps before the storm, you can be sure that you have the knowledge, supplies, and plan to protect the ones you love. Don't let a storm ruin your spring and summer, take those steps now.

Essay 6-Propane Gas: A Killer

Introduction

There are few things in this world as devastating to a family as the loss of a child. Having the bonds of life broken through disease or some natural disaster is painful enough, but to realize that your precious child was lost to something as preventable as substance abuse is almost unbearable. One small aspect of this national problem is the abuse of inhalants or "huffing."

The Dangers

The National Institute on Drug Abuse reports that about 20% of all eighth-graders have reported abusing inhalants, but only 15% of twelfth-graders responded to ever abusing inhalants. So how do you explain the fact that more eighth-graders have abused inhalants than twelfth-graders? It may be that inhalants have a devastating impact on the brain and frequent abusers are more likely to drop out of school than non-abusers. Or the abuser may have succumbed to the lethal properties of the substance being abused. Or the number of users may be growing.

In the Personal Safety section of the December 1996 edition of Professional Safety, Henry G. Wickes' article entitled "Inhaling Helium: Party Fun or Deadly Menace." brought to light the hazards associated with inhaling helium for fun. It can also remind us of the cold reality of the hazards created by consumers who misuse products for recreation (Wickes, 1996).

Two common gas products that are easily abused and can cause injuries are propane, which is contained in small metal cylinders and is used in lanterns or camping stoves, or butane found in disposable lighters. Propane bottles are filled with one to twenty pounds of propane, and butane can be found in lighters or in refill bottles of various sizes. These products are sold in hardware, grocery and discount stores, or almost anywhere outdoor sporting goods are found. The products are relatively inexpensive and may be purchased without restriction by persons of

25

nearly any age. The age group most at risk is 12-25-year-olds; however, it is reported that six percent of all children in the United States have abused an inhalant by the time they reach the fourth grade, or are about 9 years of age.

Propane and butane both react inside the body the same way when inhaled; they provide the abuser with a "high" by depressing the Central Nervous System. Symptoms include alcohol-like effects, slurred speech, lightheadedness, lack of coordination, and possible hallucinations and delusions. This state of euphoria lasts for only a few minutes. The real problem comes when an abuser attempts to continue the "high" through prolonged use. This practice can lead to asphyxiation and death. What makes this even scarier is that the product is often inhaled as a group of 2-3 people sitting around in a circle taking turns inhaling the gas as it is discharged from the container. These types of bottles have an opening controlled by a stem mechanism. By pushing down on the stem, the product is released and with the abuser's head 4-6 inches from the bottle, they simply breathe in the gas.

There are two basic hazards with this practice. First is the damage done to the body due to the propane in the nose, mouth, sinuses, esophagus, and lungs. The amount of damage will normally be tied to the amount of product inhaled and the number of times inhaled in a day, week, or month. A warning label is attached to many of the consumer sized propane bottles warning of long term effects of propane such as it being a cancer-causing agent, causing birth defects, and harming the reproductive organs. The second hazard and the most dangerous is the fire hazard.

As the abusers are discharging the gas into the air, not all the gas can be inhaled. Some escapes into the environment. In addition, the air that is exhaled contains gas causing more gas to reenter the environment. This environment may be that of a room, an auto, or a hideaway the abuser is using. The gas can create an explosive hazard in and near the area where the act is being performed. A spark created by an electrical outlet, appliance, or perhaps a cigarette may ignite this gas causing a local explosion.

This explosion will be limited in scope to the amount of gas discharged into the atmosphere of the surrounding area. The explosion may even include a flash fire that is extinguished when the gas product is consumed; however, burns would certainly occur to the face, neck, and hands, and most likely catch the clothing on fire. The type and flammability of clothing would be a factor in the severity of burns a person would suffer. Synthetic material may burn deeper into the skin causing even more severe injuries.

Even though the explosion or local flash fire is short the clothing, hair, skin, furniture, curtains, etc. can catch fire and burn on their own. These secondary fires must be extinguished, but most abusers won't think about having a fire extinguisher on hand before they start "huffing". In serious cases, the fire burns the interior of the nose, sinuses, mouth, esophagus, and lungs depending on where the gas is in the respiratory system at the time the ignition source reaches the gas.

Actual Case

One case in point is an event that involved three young men in their late teens and early twenties. These three men were fellow employees sitting inside the enclosed back of a truck sniffing the Propane that was supposed to be used for lanterns on the job site. The group had been at this for some time when one individual decided he needed a cigarette. As he lit the cigarette the back of the truck exploded in a flash fire. The fire ignited the young men's clothing and caused severe burns to their bodies. Other workmen in the area responded to the scene and put out the fire, and saved their lives. These young men were all taken to the hospital with first and second-degree burns to their bodies and two of the young men suffered second-degree burns to the inside of their mouths and throats. This one event not only cost an employer the loss of a truck and its contents, but it also cost several lost workdays, several days in the hospital, and many supervisory work hours dealing with the incident.

Summary

Why would a person take such a risk for a little enjoyment or "high"? This is a question that has plagued mankind for some time. Could it be that the abuser does not understand the dangers involved with this product? Perhaps this is one causal factor. The fact that you can buy these products off the shelf may also give the buyer a false sense of security. There are also several legal issues here. The person committing this act may be held legally responsible for any injuries or property damage resulting from the act. The other end of this spectrum is when the abuser is injured because of the misuse of this product and sues the manufacturer or seller.

Any product can be misused if someone chooses to do so. The misuse addressed in this article is intentional and the results are accidental only in that the users did not intend for a fire or explosion to occur; however, this should not relieve this person from the liability of their actions. Fire, explosion, and personal injuries are the likely results for the misuse and abuse of this type of product in the short term. Damage to the Central Nervous System, cancers, and reproductive organ disorders are the long-term legacy "huffers" should look forward to.

What can be done? Campaigns already exists that provide facts, hazards, and possible end results for abusers, and literature is available from youth centers, doctors, school counselors, and from school drug prevention programs. This is a case where we must care enough to get the facts ourselves and publicize the hazards so that potential abusers will know the result of misusing this type of product. Parents of pre-teens through young adults should be especially educated so they may be alert for the signs of abuse and aware of the gas products around the home. Employers should not become complacent about watching for behavior and performance changes from their employees. The costs associated with allowing abuse to occur on the job are potentially staggering. Don't let the tragedy strike you, be aware and prepared.

Bibliography

Wickes, Henry G, "Inhaling Helium: Party Fun or Deadly Menace?" Personal Safety Section, Professional Safety, American Society of Safety Engineers, December 1996.

Essay 7-Walking and Jogging in an Obstacle Course

Introduction

A young adult walks along a local road in the early evening hours. She is walking with friends about twenty feet from a busy road. The evening is warm and the conversation light. Suddenly the young adult is run over by a car and is dead on the side of the road. Her friends' attempts to save her life are futile, her injuries too severe. The car that drove over her speeds away. The driver may not know and perhaps not care about the young adult lying dead.

The Hazards

Is this scene an exaggeration or could this happen in your community? Yes, it could happen anywhere. Whenever a motor vehicle strikes a pedestrian, whether a walker or jogger, the pedestrian loses. This does not mean that the responsibility to prevent an accident rests only with the pedestrian. It takes caution as well as action by both vehicle drivers and pedestrians to prevent these accidents.

Drivers of motor vehicles should scan the road ahead for hazards. One of those hazards is the pedestrian. This means children and adults, runners and walkers, men and women can all be a hazard. After the driver identifies the pedestrian, they must determine if they are in the path of the vehicle or if they may step into the path of the vehicle. If the answer is yes, the driver must act to avoid them. The driver has many options that include: slowing down, driving to the left part of the lane, stopping, or taking an alternate route.

Pedestrians should walk or run facing traffic and scan the road ahead to identify hazards. One of the hazards is the motor vehicle. This means cars, trucks, and even motorcycles can be a hazard. After the pedestrian identifies the vehicle, they determine if it will be a hazard to them. Lastly, if the pedestrian determines the vehicle will affect them, the pedestrian acts to get out of the path of the motor vehicle. The pedestrian also has options. These include: changing their path, walking

29

a greater distance from the road surface, or finally jumping out of the path of the vehicle.

When it comes to traffic control devices, like the crosswalk, drivers do not have options. Drivers must stop at crosswalks if a pedestrian is attempting to cross the street or road or is in the crosswalk. Pedestrians must use crosswalks when they are provided. When a crosswalk is not provided, the pedestrian should cross the road at a right angle, quickly, and only when traffic is not coming. Many times, a driver might signal a pedestrian to cross at an area where a crosswalk is not provided. This can be dangerous. The pedestrian should use caution when a driver signals. They should not expect a driver coming from the other direction to stop. They should also be prepared for drivers to drive around the stopped vehicle.

Pedestrians should use sidewalks when provided. When they are not provided, they should walk or run on the side of the road facing traffic. This allows them to see the hazards as they approach and provides an opportunity to avoid the motor vehicle. Pedestrians should never wear earphones to listen to music, radio, or books. This will prevent them from hearing noises that may alert them to a hazard.

Summary

Pedestrians run the risk of being injured in an accident with a motor vehicle. However, it does not have to be that way. If drivers and pedestrians work together to identify and control hazards, there need not be any accidents or injuries. All communities would be much better places to live if drivers drove vehicles slower and showed a little more concern for those around them. Giving a pedestrian a break won't cost a lot of time, but it can prevent an accident. This can add up to making a safer place for pedestrians.

Essay 8-Family Homes Can Be Dangerous

Introduction

The average American home can be a dangerous place to live. This is because of the hazards we create and allow to exist in our homes. The leading causes of accidental deaths are: failure to identify hazards, underestimating personal risk, and overestimating personal ability. More than 50% of all disabling injuries occur during off work hours. Where do most of us spend our off-work hours? For many of us that is our home.

Dangers

The fact that your home could catch fire is a very scary thought for many of us. However, it is a very real consideration. Children playing with fire causes many home fires. In our homes, we should use child-resistant lighters, store matches and lighters up high, preferably locked-up, and never let children play with matches or lighters. Our homes are also at risk of an electrical fire. Many of these fires are caused by our own neglect. Each of us should not overload extension cords, replace fraying or overheating cords, and use proper size fuses in circuits. We should never run electrical cords under rugs, through walls, or through door or window openings. Each of these can damage the cord and we probably would not even know it.

At one time or another, most of us will have small children at home and they can be most at risk for an accident. Many children drown inside the home. The bathtub is the main cause. Never leave a child unattended in or near a tub of water. Many parents or adults leave the room for what appear to be safe and logical reasons. Those reasons might include: answering the door, answering the phone, or perhaps responding to another child. Another cause of drowning in children is a pail of cleaning water. Children fall into the pail while parents are not paying attention. Many children die not of drowning but from chemical pneumonia from the cleaning solvents in the water.

31

There are many children injured in the home by appliances. The cooking stove in the kitchen is one that parents usually place off-limits to their children. Parents do this to prevent the child from pulling a pan of hot material off the stove onto them. However, the sides and front of the stove may also be hazardous. Most stoves will have a hot outside to them while in use. This surface can be more than 140 degrees. That means that in addition to the tradition of keeping children away from the pots and pans on top of the stove they need to stay away from the sides and front.

Home remodeling can also create hazards for our families. Whenever possible, use latex based paints and stains. Be sure the painted area is well ventilated until completely dry. When painting, keep all oil-based paints and stains away from heat and open flame. Store paints, cleaners, and solvents outside the home in a fireproof container and keep all painting materials out of the reach of children.

You should also use appropriate clothing, shoes, and gloves for the task you are doing. This will include proper hearing, seeing, and breathing protection. You should also carefully inspect all tools and equipment before using them. Read and follow instructions on paints, solvents, glues, and other chemicals and materials. Inspect ladders before using them and position ladders correctly. An extension ladder should be used at an angle with the bottom about ¼ of the height away from the wall. Stepladders should be used on a flat level surface and you should never use the top step. Dispose of all waste properly, safely, and quickly. Don't leave it around for children or animals to get into. Plan and take your time - haste still makes waste when remodeling or repairing your home.

Summary

To keep your home safe, take a few minutes and ask yourself some questions. What will be the next accident in my home? Who will it likely involve? What should I be doing to <u>prevent</u> it? Take the answers you get from these questions and prevent future accidents. The lives and well-being of you and your family are at stake.

Essay 9-Space Heater Safety

Introduction

Each winter personnel are injured by improperly using space heaters. The key to preventing these injuries is a program that identifies the hazards associated with the specific space heater being used and implementing procedures to reduce or eliminate the hazards. This program is only effective if the procedures identified to reduce and eliminate hazards are implemented. As winter gets closer, there will be more places that will be identified as needing heat or supplemental heat. The solution for many of those areas will be space heaters. These heaters are purchased at local hardware stores. They are easy to purchase, but not as easy to use as we often assume. This essay will highlight some of the hazards with operating space heaters and recommend solutions to eliminate some of the hazards.

Operations

The improper operation of the space heater is normally the root of the problem. Proper operation begins with the identification of the individual who will operate the space heater, followed by heater-specific training, resulting in certifying of the individual. There must be specific guidance on the selection, training, and certifying of heater operators. The actual training is developed by organizational personnel using the appropriate instruction manual for the heater and is documented in the form of lesson plans. A hands-on performance evaluation is the best way to determine the skill level prior to certification. Table 1 lists the most common contributing factors for space heater accidents. As you read through the list, it becomes apparent that if the operator is properly selected, trained, and certified, many of these contributing factors can be avoided. The addition of proper supervision to the program will not only ensure the operator is trained and certified, but will identify and correct hazards with the space heater, its fuel, and use.

33

- Personnel not trained to operate space heaters.
- Using the wrong type of fuel.
- Clothing, material, and other items placed too close to the heater.
- Storage of flammable liquids near the heater.
- Heaters being refueled before allowing enough time to cool down.
- Heaters left unattended.
- Lack of or improper maintenance of the heater unit.
- Emergency procedures not established and firefighting equipment not available or non-operational.

Table 1 - Contributing Factors to Space Heater Accidents

Table 2 provides a list of control measures that supervisors can use to control or eliminate the hazards.

- Train personnel to operate space heaters safely.
- Ensure the proper type of fuel is used. Refer to the user or technical manual for recommended fuel type.
- Clothes, wet or dry, should be kept at least 24 inches away from any heater.
- Space heaters should be placed on a noncombustible base no smaller than 36 by 36 inches. The heater should be installed at least 4 feet from walls. Anything near the heater should be kept away from openings.
- Combustible, corrosive materials, and explosives should never be stored in rooms with space heaters.
- Space heaters should be turned off and allowed to cool for at least 30 minutes before refueling.
- Space heaters (including fuel lines and connections) should be checked daily for leaks and malfunctions. Only qualified personnel should make repairs. Equipment should be repaired as needed.
- Install and maintain smoke, fire, and CO2 detectors. A fire extinguisher should be available at selected fire points. At least one 10-pound dry chemical fire extinguisher is recommended. All fire extinguishers must be maintained in operating condition.

Table 2 - Control Measures to Reduce the Potential for Space Heater Accidents

There are various types of space heaters used. Space heaters that use fossil fuels should always vent their exhaust outside the space being used or potentially being used by personnel. This limits the exposure of personnel in the heated space to Carbon Monoxide gas. This gas is colorless and tasteless and normally cannot be detected without an alarm or measuring device. It is key to limit or prevent exposure to this gas.

Risk Management should also be a part of the program. This process can assist operators and supervisors in identifying and controlling hazards. It is best implemented by leaders ensuring control measures are used and that hazard and control measure information is passed to personnel who will be exposed to the hazards of operating space heaters.

Through the proper application of a space heater safety program, supervisors can control and eliminate the hazards faced by users. This program will prevent burns or deaths to users and the fire damage that often destroys equipment.

Summary

With winter fast approaching, now is the time to develop and implement a space heater safety program. This winter personnel will be injured by improperly using space heaters, don't let them be in your organization. The key to preventing these injuries is a program that identifies the hazards associated with the specific space heater being used and the implementation of procedures that will reduce or eliminate the hazards. This program is only effective if the procedures identified to reduce and eliminate hazards are implemented. Identify locations where heat or supplemental heat will be needed now, before the cold sets in. The solution for many of those areas will be space heaters. Purchase UL-approved heaters at local hardware stores and train personnel to use them. This essay highlighted some of the hazards of operating space heaters and recommended solutions to eliminate some of the hazards.

Note: This article was originally published in the Perspectives Newsletter, Volume 10, Number 1, 2010 by the American Society of Safety Engineers, Council on Practices and Standards.

Essay 10-Hazards of Crystalline Silica

Introduction

Silica, a common mineral in the earth's crust, is a major component of sand, rock, and mineral ores. Breathing it is a serious health concern. It can cause scar tissue to form in the lungs, reduce the lungs' ability to extract oxygen from the air, and cause a disease known as silicosis.

Types of Silicosis

There are three types of silicosis: chronic, accelerated, and acute. The most common type - chronic silicosis - is a unique occupational disease resulting from moderate exposure to crystalline silica over a long period of time (10 years or more). The disease can be progressive, disabling, and can lead to death. According to the Occupational Safety and Health Administration, 1,400 people died of silicosis from 1990 through 1996 (OSHA, 1972).

Dangers

Exposure to crystalline silica may also increase the risk of developing tuberculosis and other nonmalignant respiratory diseases and contribute to renal and autoimmune respiratory diseases. In addition, the International Agency for Research on Cancer has designated crystalline silica as a "known human carcinogen."

Despite the dangers, OSHA reports that more than two million workers are regularly exposed to crystalline silica dust. For example, many construction activities - such as highway repair, masonry and concrete work, and rock drilling - expose workers to silica. For construction workers, exposure to this hazard may occur when drilling rock, cutting concrete and other masonry products, and blasting.

Engineering controls for this hazard, such as exhaust ventilation and blasting cabinets, are very cumbersome and may not be feasible. So, the next option available to supervisors is to protect workers by controlling their exposure by rotating them as much as possible. Supervisors may also consider using a respirator program, as outlined in Title 29 Code of Federal Regulation 1910.134 (29 CFR, 1998). Supervisors also should ensure that workers are informed of the following information as they relate to working with silica:

- Hazards/illnesses that may be caused by exposure
- Proper conditions/precautions for safe use or exposure
- Nature of operations that could result in exposure
- Safe work practices for its handling, use, or release
- Proper housekeeping practices
- Purpose, proper use, and limitations of respirators
- Increased risks from combining smoking and exposure

When supervisors cannot reduce exposure to silica by limiting the concentrations, they must use a program of respiratory protection to protect every worker who is exposed. This process begins with identifying workers who will need to use a respirator and getting them a physical to ensure that they can properly wear one. Workers are then fitted with an appropriate respirator and trained on its use and maintenance. A Safety, Health, and Environmental specialist assigned to the organization can assist in these steps.

When workers have been exposed to dust that contains silica, their clothing or coveralls should be vacuumed. They should not clean their clothes by blowing or shaking them. In addition to clothing, exposed surfaces should be kept free of silica dust. If the dust on these surfaces is disturbed, it could become airborne and breathable. Dry sweeping and using compressed air for cleaning floors and other surfaces should be prohibited. When vacuuming is used, the exhaust air from the vacuum should be filtered. Gently washing the surfaces is preferred, when practical. All food, beverages, tobacco products, nonfood chewing products, and unapplied cosmetics should not be used in work areas with silica dust. Workers should also be able to wash their hands with soap and water after exposure.

Summary

This essay provided supervisors with the basic knowledge to make informed decisions concerning steps necessary to prevent or control exposure to the hazards of working with crystalline silica. Although supervisors cannot control all the hazards that workers face, controlling this one could prevent needless illness now and in the future.

Bibliography

Occupational Safety and Health Administration Instruction CPL 2-2.7, October 30, 1972.

Title 29 Code of Federal Regulation 1910, Section 134, "Respiratory Protection," April 23, 1998.

Reference

Army Regulation 11-34, The Army Respiratory Protection Program, February 15, 1990.

Occupational Safety and Health Administration Semiannual Regulatory Agenda, 68:73196-73228, Sequence Number 90 in part 11,1218-AB70-2103, "Occupational Exposure to Crystalline Silica," December 22, 2003.

Note: This article was originally published in the U.S. Army's Engineer Bulletin, July-September 2004 edition.

Essay 11-Brown Recluse Spider–A Serious Hazard

Introduction

Dr. Susan Jones of the Ohio State University Extension Office writes in her fact sheet on the spider that "The brown recluse belongs to a group of spiders that is officially known as the "recluse spiders" in the genus *Loxosceles* (pronounced lox-sos-a-leez). These spiders are also commonly referred to as "fiddle back" spiders or "violin" spiders because of the violin-shaped marking on the top surface of the fused head and thorax. This feature can be very faint depending on the species of recluse spider or how recently the spider has molted (Ohio State, 2004). In this essay, the author will provide a quick overview of the spider, its habits, locations and the severity of its venom. Many municipalities and public organizations are responsible for areas that may contain this spider and employees that may encounter it.

Characteristics

The common name, brown recluse spider, pertains to only one species. The name refers to its color and habits. It is a reclusive creature that seeks and prefers seclusion. There is a dark violin marking on the spider's back.
This spider is in a limited portion of the United States, which includes Missouri, Arkansas, Mississippi, Louisiana, and Alabama, as well as parts of Texas, Oklahoma, Nebraska, Illinois, Indiana, Kentucky, Tennessee, North Carolina, and Georgia. The spider can be found in dark, damp areas. A common area is outside restrooms at campsites or picnic areas. The spider is small, about 3/8 inches long, and 3/16 inches wide (Ohio State, 2004).

The Bite

Dr. Jones of the Ohio State Extension Office describes how the "physical reaction to a brown recluse spider bite depends on the amount of venom injected and an individual's sensitivity to it. Some people are unaffected by a bite, whereas others experience immediate or delayed effects as the venom kills the tissue at the site of the bite. Many brown recluse bites cause just a little red mark that heals without event" (Ohio State, 2004).

Dr. Jones even goes on to explain that "initially, the bite may feel like a pinprick or go unnoticed. Some victims may not be aware of the bite for 2 to 8 hours. Others feel a stinging sensation followed by intense pain. The bite of the brown recluse spider can result in a painful, deep wound that takes a long time to heal. Fatalities are extremely rare, but bites are most dangerous to young children, the elderly, and those in poor physical health. When there is a severe reaction to the bite, the site can erupt into a hole in the flesh due to damaged tissue. The open wound can range from the size of an adult's thumbnail to the span of a hand. The dead tissue gradually comes off, exposing underlying tissues. The sunken, ulcerating sore may heal slowly up to 6 to 8 weeks. Full recovery may take several months and scarring may remain" (Ohio State, 2004).

Richard Vetter wrote in an article for the Dermatology Online Journal, "Recluses typically bite when they are trapped between flesh and another surface, as when a sleeping human rolls over on a prowling spider, or when putting on clothing or shoes containing spiders" (Treatment, 2008).

Treatment

For severe bites, doctors often provide rigorous treatment over a six-week period. Fortunately, most bites from this spider are relatively minor, requiring little more treatment than an average spider bite; however, this spider can and does produce severe reactions in some victims. This spider has its own website which covers all aspects of the spider. This website lists the following first-aid measures:

- call 911,
- use ice over the injury to lessen the pain and swelling,
- when possible, place the bitten body member such as arm or leg above the heart,
- wash whole body area with cool water and soap,
- stay calm - nervousness helps venom spread, and
- put sterilized bandage on the injury and use ibuprofen or similar anti-inflammatory drugs for pain relief. (Treatment, 2008)

Prevention

Richard Vetter also wrote in his article for the Dermatology Online Journal about ways to reduce bite risk from recluse spiders that include:

- Keep beds away from walls; remove bed skirts and items under the bed so that the only pathway to the bed is up the legs.
- Keep clothing off the floor; if it is on the floor, shake it vigorously before dressing.
- Store all intermittently used items such as gardening clothing, baseball mitts or roller skates in completely enclosed spider-proof boxes or bags."

When public or contract maintenance workers are working on equipment or cleaning parks, camping areas, and playgrounds, they can encounter this spider. This can be avoided by wearing gloves and long sleeve shirts, raising items on the ground with care, clearing areas before sitting on the ground, and knowing what the spider looks like. It is also important for workers to have a first-aid kit readily available or a phone to contact help when needed.

Summary

The brown recluse isn't your everyday spider. Although most bites are minor and result only in discomfort, some bites can lead to serious wounds requiring hospitalization and treatment for several weeks. In this essay, the author provided a quick overview of the spider, its habits and locations, and addressed the severity of its venom.

Bibliography

Identifying and Misidentifying the Brown Recluse Spider, Rick Vetter, Dermatology Online Journal, University of California Riverside, 1999. Retrieved from http://dermatology.cdlib.org/DOJvol5num2/special/recluse.html on September 30, 2008.

Ohio State University Fact Sheet, Brown Recluse Spider, Susan C. Jones, Ph.D., 2004. Retrieved from http://ohioline.osu.edu/hyg-fact/2000/2061.html on September 30, 2008.

Treatment, The Official Brown Recluse Spider Web Site. Retrieved from http://www.brownreclusespider.org/brown-recluse-spider-treatment.htm on September 30, 2008.

Note: This article was originally published in the Perspectives Newsletter, Volume 8 Number 3 in 2009 by the American Society of Safety Engineers, Council on Practices and Standards.

Essay 12-Recognizing and Controlling Human Factors

Introduction

Humans by omission or commission cause or contribute to incidents that result in losses to property, people and resources. By eliminating or controlling this human factor one can eliminate or control the potential for accidents.

Safety, Health, and Environment programs play an important role in preventing the injuries and illnesses experienced by employees. Accidents can be reduced by leveraging human resources management principles to control human factors. These principles consist of hiring the right person, ensuring potential employees are technically competent and physically capable, bringing new employees on board correctly, and managing aspects of the new employee's career with the organization.

This essay will highlight principles of human resources management that, if used, can provide management with the means to control and eliminate human factors and the potential for accidents.

Supervisors Role

The supervisor plays a central role in the hiring process. Supervisors must make an up-front investment by taking the time to critically consider and then effectively communicate to the Human Resources Office the critical knowledge, skills, and abilities required by a job as identified by a job analysis. This begins with a thoughtful conversation in which the supervisor furnishes information that will enable a human resources specialist to more effectively design a job, market the vacancy, and assess applicants. The supervisor again plays a key role when conducting structured interviews as the last step in the assessment process. The supervisor then makes the hiring decision. Later the supervisor must ensure the new employee is brought into the new job properly to ensure he or she knows about required policies and procedures, use of equipment, reporting hazards and accidents, and how to identify training and growth needs.

Human Resources Specialist Role

The human resources specialist plays a supporting role in the hiring process. They participate with supervisors in the up-front investment by taking the time to critically consider and then ensure they understand the critical knowledge, skills, and abilities required by a job. It is at this point that the human resources specialist takes the supervisor furnished information and designs the job, markets the vacancy, and assesses applicants.

Safety, Health, and Environmental Specialist (SHE) Role

The SHE Specialist also plays a supporting role in the hiring process. They provide information about hazards associated with a job, extreme weather exposure, Ergonomic issues, and any special safety skills that the employee may need. The SHE specialist then works with the human resource specialist to develop a job hazard analysis.

Job Analysis

Job analysis is the foundation of recruiting human resources and is vital to selecting the proper employee. Identifying the best person for the job requires that the supervisor fully understands essential duties of the job and the environment in which the job will be performed. By conducting a job analysis, the supervisor systematically identifies the knowledge, skills, and abilities necessary for success on the job. Then the supervisor, with the help of a human resources specialist, develops valid and effective selection tools.

The supervisor need not conduct a job analysis every time he or she wants to fill a job vacancy; however, it is essential the first time a job is created and filled. If job openings within the same occupation occur frequently, the supervisor may rely on selection tools developed from recent job analyses of that occupation. These selection tools include structured interviews, written tests, use of an assessment center, and work samples.

Whether a job analysis should be conducted for a job depends on whether the job is new and if not new how current the existing job analysis is. Periodic review of job analyses is important. If the requirements of the job frequently change the supervisor should review the job analysis prior to each job being filled to ensure the selection tools are still valid. Contrasting that with the requirements of a job that changes very little. The job analysis may need to be reviewed less frequently; perhaps annually regardless of how many times a job is filled during that time.

Job Hazard Analysis

A job hazard analysis is conducted by a SHE professional. The job hazard analysis focuses on job tasks to identify hazards before they occur. In contrast a job analysis has a much broader scope of looking at what the job entails. Like the job analysis the job hazard analysis includes the relationships that exist among the worker, task, tool, and the work environment, but does not focus on it. The job analysis must precede the job hazard analysis because it facilitates the identification and control of hazards associated with the job that and allow hazards to be identified in the job description for consideration.

Physical Requirements

It is important to identify the physical requirements of the job up front. These include walking, standing, climbing stairs and ladders, and lifting. Noise and visual effort is also important. Another piece of information that is important is how frequent the requirements are experienced. This should be noted as sustained, intermittent, or seldom. The results of this information will determine the requirements for a pre-employment physical. This is an important part of any hiring process. Without a proper pre-employment physical, the employer accepts responsibility for the employee as is. This could result in significant costs if an employee's previous condition is exacerbated by the current job.

Environmental Requirements

This involves noting the environment the job will be performed in. This includes heat, cold, height, underground, in the dark, in inclement weather, etc. This information will allow the supervisor to also identify protective equipment and clothing to reduce the risk to the employee from the specific environmental requirements. This information will also support the pre-employment physical in such areas as cold and hot weather injuries where a previous injury may leave an employee susceptible to future injuries.

Position Classification

In the public sector the next step after the Job Analysis is to have the job classified. This is called Position Classification and describes a process through which jobs are assigned to a pay system, series, title, and pay grade, based on a consistent application of position classification standards. Positions are classified to achieve uniformity and equity using a common reference across organizations, locations, and agencies. Classification standards cover one or more occupations, usually including a description of the work performed; official titles; and criteria for determining grades. Most human resources offices have developed grading guidance, broad standards that serve as functional guides, and provide criteria for determining the pay level of work. These can normally be obtained from the human resources office serving the supervisors organization. Position classification standards and guidance also distinguish between white collar and blue collar (trades, craft, and labor) jobs. It is easy to confuse classification with qualifications; however, each has its own distinct purpose.

Classification pertains to a specific job and the evaluation process that determines the appropriate pay system, occupational series, title, and grade. While qualifications pertain to a person who is applying for or encumbers a job and describes the knowledge, skills, and abilities a person must have to be successful. Normally there are three kinds of competencies.

- Required Competencies-required by the position.
- Enabling Competencies-not required by the position, but are instrumental in assisting the incumbent employee to perform the job successfully. An incumbent should work to develop these competencies.
- Developmental Competencies-not required by the position, but necessary for the incumbent to move up to the next pay grade.

Job Descriptions

A job description is a statement of the major duties, responsibilities, and supervisory relationships of a position. Simply put a job description indicates the work to be performed by the employee incumbent to the position. The purpose of a job description is to document the major duties and responsibilities of a position. The job description cannot spell out in detail every possible activity during the work day. Human resources offices often maintain a library of job descriptions. An outline of a description is shown in Figure 1.

Title (series and pay grade)
Office or Work Organization

INTRODUCTION

MAJOR DUTIES AND RESPONSIBILITIES

FACTORS

1. KNOWLEDGE REQUIRED BY THE POSITION

2. SUPERVISORY CONTROLS

3. GUIDELINES

4. COMPLEXITY

5. SCOPE AND EFFECT

6. PERSONAL CONTACTS

7. PURPOSE OF CONTACTS

8. PHYSICAL DEMANDS

9. WORK ENVIRONMENT

Figure 1 – Outline of Job Description

Summary

With Safety, Health, and Environment playing such an important role in the work life of Americans, it is important to use all the tools available to identify hazards before they cause an accident. One tool is to use the processes in place to focus on human factors that are often involved in accidents. This essay highlighted principles of human resources management that, if used, can provide management with the means to control and eliminate human factors and the potential for accidents.

Essay 13-Motorcycle Risk Management

Introduction

Can Risk Management be the solution to everything? Maybe not, but it can provide a great option to fatal motorcycle accidents. Who says so? The Motorcycle Safety Foundation (MSF) says so. They have done a great deal of work to develop training programs to provide riders with the skills necessary to prevent motorcycle accidents. Riding a motorcycle can be a very dangerous sport or activity, but with proper training and application of Risk Management, riders can act to prevent crashes. Many states use the MSF curriculum for motorcycle rider training and many more should.

History

In the April 25, 2003 issue of USA Today, Jayne O'Donnell reported that motorcycle fatalities were up in 2002 for the fifth straight year. She indicated an increase of 3%. Ms. O'Donnell obtained her information from the National Highway Traffic Safety Administration (O'Donnell, 2003). The National Highway Traffic Safety Administration website has two very interesting reports on this subject. In an article in the April 2003 Countermeasure publication, Master Sergeant Dave Hembroff raised the issue of motorcycle rider's risk of being involved in an accident. He indicates in his article that a rider who has not taken a rider training course is nine times more likely to be involved in an injury accident. From October 2002 through February of 2003, soldiers of the Army had eighteen motorcycle accidents for the fiscal year. Six soldiers died in those accidents (Hembroff, 2003).

Conducting Risk Management

Accidents are normally the result of a series of events or factors that lead up to the accident. By controlling or eliminating those factors, a rider reduces the risk of being involved in a motorcycle accident. There are three primary areas that should be addressed in conducting risk management for motorcycle riding. These three areas are: rider factors, motorcycle factors, and road and traffic factors; see Table 1 for additional information. Each of these areas contains several elements that determine the rider's risk of being involved in an accident.

Factor	Explanation
Rider Factors	Experience, Training, Protective Clothing and Equipment, Consumption of alcohol and drugs, or lack of sleep
Motorcycle Factors	Size and Fit, Working Condition
Road and Traffic Factors	Road and highway conditions, weather conditions, traffic conditions

Table 1 - Three Factors to Consider in Risk Management

Riders should always be prepared to ride the motorcycle. That may sound a little strange, yet it is true. The rider of a motorcycle must focus his or her attention on the task of riding the motorcycle as well as the actions of other drivers, wildlife, and the condition of the road all at once. This is far more focus than any automobile driver puts into the task of driving. The amount of time a rider has on their motorcycle has a great impact on the potential for an accident. Normally, the more you ride, the better rider you become. Army personnel are required to complete a motorcycle rider course to ride a motorcycle. This course is offered at most installations and provides basic information about riding. Many other organizations should provide this training. The more training you get, the better a rider you will become. Go to www.msf-usa.org to get more information about motorcycle rider courses in your area.

There are items of protective clothing and equipment that each rider must wear; see Table 2 for a complete list. Many riders wear at least an approved half shell helmet. However, if they were to wear an approved ¾ shell or a full-face, they could reduce their risk even more. The same thing goes for the shirt or pants. The rider can use a regular pair of pants with a long sleeve shirt and get by. Safer still would be to wear the new jackets and pants with ballistic protection sold by many manufacturers today. This ballistic protection is in areas where the body is injured in a crash. Using it will greatly reduce the risk of injury in an accident.

Clothing and Equipment	Explanation
Helmet	Helmets come in full-face, ¾ shell, and ½ shell. The Department of Transportation or Schnell Foundation must approve the helmet. The full face provides the best protection followed by the ¾ shell. Half shells provide the least amount of protection.
Gloves	Gloves should be leather and have full fingers. It is best to purchase motorcycle gloves because they are sewn to put the seams outside the glove and curve the fingers.
Shirt	Long sleeve made of durable fabric. Consider a jacket or riding suit with ballistic protection.
Pants	Long legged made of durable fabric. Consider pants or riding suit with ballistic protection.
Shoes	Over the ankle boots or shoes. No high heels on boots and no large toes. Consider a pair of motorcycle boots.
Protective Eye Wear	Don't rely on the face shield to protect you. Wear impact resistant eyewear even if you wear a face shield. Invest in a pair of impact resistant sunglasses.

Table 2 - Required protective clothing and equipment

Since riding a motorcycle takes a great deal of concentration, it is surprising that many riders drink and ride. Yet it still happens. If you plan to drink, don't ride. Your chances of having an accident are far greater. Along with drinking, riders should make sure they don't take prescription or over the counter medications that may cause drowsiness prior to riding. Read the label, and if it has a warning about driving or operating heavy equipment or machinery, that means you don't ride. Along with these hazards comes the risk of riding when tired. It is very hard to drive a car when tired; it is only worse trying to ride a motorcycle when tired. You may feel like you are riding fine until an emergency occurs and you can't properly react to it.

Even if you are prepared to ride, is your bike ready to be ridden? First, does it fit you, and secondly, is it in good working order? Is your bike the right size? You can tell by sitting on the seat and putting both feet flat on the ground. If you can't do this, the bike is too tall. Now try to reach all the controls. You must be able to reach the handlebars, clutch lever, brake lever and pedal, throttle, and shift lever with ease. If you can't reach all the controls, have them adjusted. Now is your bike in good working order? How do you know? The MSF has a pre-ride check that is represented by the acronym T-CLOCS. This represents T=tires and wheels, C=controls, L=lights and other electrical items, O=oil, C=chassis, and S=side stand. By conducting a quick inspection and fixing those items that don't work, you can greatly reduce your risk.

The last things to consider are the road and traffic conditions. You can choose the place and time you ride so make it the safest time and place. Don't ride in areas with limited visibility or rough and sandy roads. These can cause or contribute to an accident. You may also want to avoid heavy traffic times. Most car and truck drivers are not watching for motorcyclists and often don't see them. Not riding in these situations can reduce your risk.

Strategy

In addition to identifying the hazards and eliminating those you can prior to riding, the MSF recommends a strategy for riding your motorcycle. The strategy is known by the acronym SEE. S is to search for hazards constantly as you ride. E is to evaluate those hazards first to determine if they have an impact on you and then to develop a course of action for each. The second E is to execute the course of action you determined in the evaluation step. Sounds familiar, doesn't it? This is a constant update of the risk management process. The more you use it, the better you will become.

Summary

Whether you are a new rider or a 20-year veteran, you can become the victim of a motorcycle accident. You can reduce the potential for that accident by using the risk management process explained in this essay to identify and eliminate hazards. Don't become overwhelmed with all the hazards. Riding a motorcycle is more dangerous than driving a car and most, if not all riders know this. To be successful, control the hazards you can and reduce your risk. Let motorcycling be fun and enjoyable.

Bibliography

O'Donnell, Jayne. Traffic deaths rise to 12-year high, USA Today, April 25, 2003, page A-1.

Hembroff, Master Sergeant Dave, Learn and Live, Countermeasure, April 2003: 16-18.

Note: This essay was published in the July-September 2003 edition of the United States Army's Engineer Professional Bulletin where it was recognized by "Silver Quill Award."

This essay was also published in the IMPAX Magazine. Volume 1, Issue 2, March/April 2005 by the United States Army Safety Center.

Essay 14-Using Collateral Duty Safety Representatives

Introduction

Many organizations use collateral duty safety representatives. Furthermore, as more organizations downsize they may switch from full-time safety personnel to collateral duty safety representatives. The key to making this decision successfully is to implement a logical process to select, train, and support those appointed. This essay will look at the best way to do just that.

The decision to use collateral duty safety representatives alone or to supplement a full-time staff should be based on the best way to conduct a safety program. The decision should include cost, professional standards, and return on investment. The author's experience is that collateral safety representatives are being used and many are not properly selected, trained, or supported. If their use increases this trend will continue or perhaps increase.

Organizational Responsibilities

The Occupational Safety and Health Act of 1970 gives the requirement to prevent needless accidents and injuries in the workplace. Section 5 (a) (1) says that "each employer shall furnish to each employee employment and a place of employment which are free from recognized hazards that are causing or are likely to cause death or serious physical harm to his employees" (OSHA, 1970). Section 19 of the Act says that federal agency heads will implement a safety and health program. This act does not apply to state or municipal governments. However, nearly every state has its own standard covering public employees. This means that public organizations must prevent accidents by recognizing and eliminating hazards in the workplace. How the program is staffed is not as important as its effectiveness in preventing accidents.

Resources

The public manager and his or her human resources manager must identify resource needs and justify them in the budget processes. The following ratios will aid in the budget decision making process and are based on the author's experience. Collateral duty safety representatives should be used at a ratio of one collateral safety representative for every five hundred employees. This assumes 8 hours a week on safety duties. With employees that work three shifts the collateral duty safety representative ratio should be one per shift instead of the number of employees. If full-time safety personnel are used the ratio should be one for every 3, 500 employees. If a mixture of collateral and full-time safety personnel are used the ratio should be a full-time safety person at the headquarters level with collateral duty safety representatives assigned at a ratio of one collateral safety representative for every 1,000 employees at the plant or operational level (Fanning, 2003).

The safety budget must include the salaries of the collateral and full-time safety personnel. The time collateral personnel at the operational level spend on safety is time they are not doing their primary job of production or service. This cost should be equal to or less than the amount management budgeted. This cost is often overlooked. In addition to the salaries management should budget $0.24 per employee per month to purchase promotional items, safety awards, standards, regulations, and awareness material. Awareness material includes posters, flyers, stickers, pins, and banners for specific hazards in the workplace.

Selection

If management has decided to use collateral duty safety representatives the first step is to select the right person. Collateral duty safety representatives used alone or to support a full-time staff are spread throughout the organization and focus on providing services to their section. If the second duty assigned is related to safety such as engineering, environmental protection, facility management, or nursing the duties complement each other. The normal safety duties are in a pamphlet published by the American Society of Safety Engineers (Scope and Functions, 1996), see Table 1.

- Serve as management's representative on all aspects of safety.
- Interpret safety policies and procedures.
- Conduct periodic surveys and inspections.
- Conduct follow-up hazard abatement.
- Maintain records of surveys and inspections.
- Assist supervisors in investigating accidents.
- Follow-up with the director on injuries and property damage.
- Collate reports.
- Provide information to line organizations.
- Coordinate with the personnel office on the OSHA log and worker's compensation reports.
- Coordinate an early return-to-work program.

Table 1 - Major Duties of the Additional or Collateral Duty Safety Representative taken from Scope and Functions of the Professional Safety Position, American Society of Safety Engineers, form SF-10/M-M-2/96

When collateral duty safety representatives are appointed, they are seldom told what their new duties are or where to get assistance. There are a few books on the market that can help. One source of assistance is the Career Guide to the Safety Professional published jointly by the Board of Certified Safety Professionals and the American Society of Safety Engineers. The guide outlines the knowledge of a safety person on pages three and four (Career Guide, 2002), see Table 2. If possible, management should select a person with the knowledge areas outlined in Table 2. If the organization does not have someone with these knowledge areas, they should select a person who can learn them.

Chemistry	Biology	Physics	Ergonomics	Economics
Psychology	Physiology	Biomechanics	Medicine	Engineering
Sociology	Geology	Business	Management	

Table 2-Safety Knowledge Area–Career Guide to the Safety Profession, pages 3-4.

Training

Management must provide training that the collateral duty safety representative needs. To do this, management must be familiar with the collateral duty safety representative's duties, the functions of the organizational safety program, and how to assist the collateral duty safety representative in determining the training and skills they possess. Training may be provided through local colleges, OSHA approved regional training centers, or private companies. Each knowledge area in

Table 2 can be learned in a single class or several may be learned simultaneously in a single course. At Table 3 is an excerpt from the OSHA training catalog. This serves as an outline of the courses that exist to prepare personnel.

Course Number	Course Title
510	Occupational Safety and Health Standards for the Construction Industry
511	Occupational Safety and Health Standards for General Industry
1020	Basic Accident Investigation
1050	Introduction to Safety Standards for Safety Officers
1410	Inspection Techniques and Legal Aspects
2010	Hazardous Materials
2040	Machinery and Machine Guarding Standards
2070	Fire Protection and Life Safety
2200	Industrial Noise

Table 3 – Excerpts from OSHA Training Institute Course Catalog.

Support

After the individual is selected and trained, management must support them. Ray Boylston says that "managing safety and health programs is one of management's most important responsibilities" (Boylston, 1990). First, the manager meets with the collateral duty safety representative and his or her supervisor. The manager tells them where management thinks the safety program should go, how much time should be spent on safety duties, the reporting chain for safety issues, the duties of the position, and how those duties will be measured for performance evaluations. Management should supplement the individual's job description in writing with the safety duties.

Management must give the collateral duty safety representative access to senior managers. The senior manager must sign a policy memorandum and make sure that other managers and supervisors buy-in to the program. Ray Boylston also says that "the ranking manager must set the proper example by his or her actions and must demand a similar commitment from the entire line organization" (Boylston, 1990). If the program belongs to the collateral duty safety representative, it is doomed to fail. They cannot do it on their own, management must do their part.

Conclusion

Many public organizations use collateral duty safety representatives. With today's fiscal realities, many organizations may need to cut full-time personnel and use collateral duty representatives. After making this decision, management should implement a logical process to select, train, and support the person appointed. If these three areas are given proper attention, the safety program can prevent needless accidents and the cost associated with them. If you use collateral duty safety representatives in your organization, do you use a logical process to select, train, and support them?

Bibliography:

Boylston, Ray, Managing Safety and Health Programs, Van Nostrand and Reinhold, 1990.

Career Guide to the Safety Profession, Board of Certified Safety Professionals and the American Society of Safety Engineers, 2000.

Fanning, Fred, Basic Safety Administration: A Handbook for the New Safety Specialist, American Society of Safety Engineers, 1993.

Occupational Safety and Health Act, Public Law 91-596, U.S. Government Printing Office, 1970.

OSHA Training Institute Course Catalog webs site. Retrieved on December 2, 2004 from http://www.osha.gov/fso/ote/training/oti_catalog.html.

Scope and Functions of the Professional Safety Position, American Society of Safety Engineers, 1996, Form SF-10/M-M-2/96.

Note: This article was originally published in the Perspectives Newsletter, Volume 5 Number 1 in 2005 by the American Society of Safety Engineers, Council on Practices and Standards.

This article was then selected as the Public-Sector Practice Specialty newsletter article of the year and was published in the Council on Practice and Standards "Best of the Best from the 2005/06 Newsletters Article", 2006 Edition.

Essay 15-Focused on Activity or Results?

Introduction

This essay addresses a topic that the author has thought about for many years. Recent mine accidents remind us that lessons once learned are often forgotten. This essay does not present information about individuals to impugn their character, but to look at the situation in which the people found themselves and the decision they made.

On March 25, 1947, a deadly coal mine explosion rocked the calm, peaceful town of Centralia, IL. During World War II, this mine provided coal to the war effort. A charge ignited built-up coal dust and caused the explosion. This explosion should not have surprised anyone. Many public-sector safety professionals from state and federal agencies knew of the hazards because of inspections, union complaints, and letters to state officials. These same safety professionals had notified various officials of mine safety agencies and the mining company of the hazards on more than one occasion. Officers of the mine's union had also pressed for the hazards to be corrected. Failure to act to abate the hazards resulted in the loss of 111 hard-working men who had spent much of their lives mining coal.

This essay reports the circumstances that surrounded a mine explosion that killed 111 miners in a small town in Illinois. The essay will answer the question, "Should public sector safety professionals focus on activity or results?" Although this incident occurred 60 years ago, its lessons are as applicable today as they were in 1947. Knowledge of these lessons makes this essay relevant to today's safety professional.

A Brief History

In 1883, the state of Illinois was divided into five mine inspection districts, and Governor J. Hamilton appointed an inspector for each district to oversee practices and safety regulations. In that same year, a Board of Examiners for Mine Inspectors was created to judge the qualifications of candidates for the position of mine inspector. In 1899, this board's name changed to the State Mining Board. It was also charged with supervising the state mine inspection service.

The Illinois Department of Mines and Minerals was created in 1917. It absorbed several boards and commissions created earlier to regulate mining. This department carried out the policies of the State Mining Board (Illinois Secretary of State, 2003).

Coal miners went on a national strike. In response to the need for coal to support the war, President H. Truman ordered the mines seized. That same month, a letter was signed between the United Mine Workers of America (UMWA) and the Department of Interior for the department to run the mines. This letter also authorized the enforcement of federal safety rules in the mines. In 1946, the federal government developed an organization called the Coal Mines Association (CMA) to administer the mines.

The Centralia Coal Company owned Centralia Mine Number 5. The mine was opened in 1907 and covered nearly 6 square miles underground. The mine employed 250 men and produced 2,000 tons of coal each day. On March 25, 1947, the mine exploded, killing 111 miners (Barlow, 1948). In the 1947 report on the explosion, the Bureau of Mines described the reason for the explosion as "coal dust raised in the air and ignited by explosives fired in a dangerous and non-permissible manner." Heavy deposits of coal dust were present along the roadways and on the roof, ribs and timbers in working places and entries (Bureau of Mines, 1947).

According to the Centers for Disease Control (CDC), from 1839 to 2001, 717 mine disasters occurred in the US, killing 15,196 people. CDC ranked the Centralia Mine No. 5 disaster as 23rd worst in the nation (CDC, 2001). The U.S. Mine Rescue Association (USMRA) lists the Centralia disaster as one of the five worst coal mine disasters since 1940 (USMRA, 2003).

Events Leading Up to the Explosion

In 1941, Illinois Governor D. Green appointed R. Medill, who worked on the governor's campaign, as director of the Illinois Department of Mines and Minerals. In 1941, the governor also appointed D. Scanlan, recommended by his state representative, as one of the state's 16 mine inspectors (Martin, 1948).

Scanlan was the inspector of the district that included Centralia Mine Number 5. Scanlan inspected the mine several times in the years before the explosion. He made a report of each inspection and sent them to the Illinois Department of Mines and Minerals. In many of these reports, he noted that the mine was highly explosive due to coal dust buildup. R. Weir, assistant director for the Illinois Department of Mines and Minerals, received and processed his reports. Weir signed a letter for each inspection that identified issues, provided recommendations and requested a response from the company. In 1942, F. Prez, a mine inspector from the U.S. Bureau of Mines, conducted a federal inspection of Centralia Mine Number 5. His findings and recommendations were like Scanlan's. Prez also noted the mine was highly explosive (Martin, 1948).

In November 1944, W. Rowekamp, the UMWA Local 52 recording secretary, sent a letter to Medill noting that the conditions of the roadways were dirty and dusty and becoming dangerous. The letter also said the company ignored Scanlan's recommendations and begged for Medill's prompt action. Weir received the letter and sent Scanlan to investigate. Scanlan conducted an inspection that identified the hazards, and N. Prudent, mine superintendent, said he would spray the roads within a week. In his December 1944 inspection, Scanlan noted that the mine was again dirty and recommended closing the mine until it was cleaned. Weir's letter on this inspection did not specify closing the mine (Stillman, 2000).

In early 1945, Scanlan phoned Medill about the hazards. Medill told him to write him a letter. In February 1945, Medill received the letter and forwarded it to W. Young of Bell and Zoller Coal and Mining Co., the parent company of Centralia Mine Number 5. Young answered that he hoped coal production would slow down in the future so he could respond to these recommendations (Stillman, 2000).

In March 1945, Scanlan told Medill that if an explosion occurred, it would spread throughout the mine and probably kill the men in it. Scanlan also claimed that Medill said they would need to take that chance. Medill denied the conversation.

In April 1945, the UMWA Local 52 filed charges against mine manager Brown for blowing a charge with the miners in the mine. In May 1945, Weir wrote to the State Mining Board stating that the company admitted to the charge, but claimed it was an emergency and would not happen again (Stillman, 2000).

In April 1945, Scanlan told Prudent that he had to shut down the mine or clean it. The mine dug coal 4 days a week and cleaned for 3 days. The conditions in the mine improved until June 1945 when Scanlan and Prez both found dust buildup. In October 1945, Scanlan wrote another letter to Medill. Based on this letter, Medill wrote another one to the company.

In December 1945, Medill called a meeting of the State Mining Board. Scanlan was on the agenda to speak to the board; however, before the board meeting, B. Schull, a member of the State Mining Board, asked Scanlan to withdraw a request he had made for sprinklers to be installed in Centralia Mine Number 5. Scanlan refused and was never called before the board.

A committee was sent to the mine to investigate charges filed by the UMWA Local 52 against Brown and Prudent. Prudent led the committee through the mine. The UMWA Local 52 was not notified of the visit nor asked any questions pertaining to the visit or to the charges. Medill notified the UMWA Local 52 that the committee found insufficient evidence for the charges the union had filed against Brown. This information was only provided to the union after a request by the union secretary (Stillman, 2000). Readers of this essay may recall that representatives of the company admitted to Weir that they had committed the charges.

In February 1946, Rowekamp wrote to Medill to tell him that UMWA Local 52 members were dissatisfied with the committee findings. A second letter was sent to the governor. A secretary to the governor sent the letter to Medill. Medill responded by explaining the situation from his perspective. The governor's secretary sent a response to the union stating that the State Mining Board would consider their request. Medill reprimanded Scanlan and directed him to cut down his reports. Medill also asked Scanlan's political sponsor if he could fire him. The sponsor refused (Stillman, 2000).

In November 1946, Prez conducted an enforceable inspection of Centralia Mine Number 5. Also in November, a CMA representative sent a letter to the mine directing management to fix all hazards identified in Prez's report and to respond to CMA in writing to confirm abatement. In November 1946, N. Niermann, the new mine superintendent, sent a letter to CMA telling them they could not comply because of a miner strike (Stillman, 2000).

In January 1947, a CMA representative wrote a second letter to Centralia Mine No. 5 directing management to correct the hazards and to reply by February 1947. A representative of Bell and Zoller Coal and Mining Co. sent a response stating that many hazards were corrected. In February 1947, a representative of CMA wrote back asking for details. In March 1947, a representative of Bell and Zoller sent the details to CMA. On March 25, 1947, the mine exploded, killing 111 miners (Stillman, 2000).

Investigation Results

The U.S. Bureau of Mines investigated the explosion. In 47 pages, the investigation team explained the circumstances of the explosion. The team found that "the explosion was localized and confined to four working sections of the mine. The two remaining sections into which the explosion failed to propagate were affected by rock dusting. The explosion failed to propagate further in every instance when it reached or as it approached the rock-dusted zones on the entries." The finding that the propagation of the explosion stopped when it reached rock-dusted zones supported the inspectors, identifying rock-dusting as the control measure for dusty conditions (Bureau of Mines, 1947).

As for an ignition source, "The investigation team believed the only possible ignition sources present at the faces or walls at the time of the explosion were the open lights of the shot firers, a few others, and the detonation of explosions." To further support the investigation team's belief that explosives ignited the coal dust, they found "indications that the top right shot in Face 1 was under burdened. This was evidenced by the shot not pulling down all the coal as in normal fashion."

Finally, the explosion could have been ignited by an improperly stemmed shot. The investigation team found "a blown out shot of explosives that was stemmed with coal dust or an under burdened shot of explosions could have ignited the coal dust that was raised by preceding shots of explosions" (Bureau of Mines, 1947).

Earlier, it was noted that in April 1945, the UMWA Local 52 made charges against the mine manager for firing explosives in an unsafe manner with miners in the mine. The State Mine Board reviewed the case, but it did not uphold the charges. The board failed to uphold the charges despite the mine manager's admission to them. Compare that incident with the investigation team's finding that "permissible explosives were being fired in a non-permissible manner with caps and fuses, and coal dust used for stemming" (Bureau of Mines, 1947).

In summary, the U.S. Bureau of Mines investigation found that the explosion originated at the face of one west entry; that it was strictly a coal dust explosion, which was propagated by coal dust throughout four working sections of the mine; and that the coal dust was raised into the air and ignited by explosives fired in a dangerous and non-permissible manner (Bureau of Mines, 1947).

This was the scenario Scanlan and Prez identified. It is unfortunate that mine management did not accept the inspector's findings and recommendations. The investigation report supported Scanlan and Prez's recommendations when the team identified that the presence of rock dust in entries, even though somewhat deficient in quantity, was the most important factor in preventing the spread of the explosion throughout the mine and to the shaft bottom (Bureau of Mines, 1947).

To clearly see what should have been known before the explosion, it is important to understand what the Federal Mine Code for Bituminous Coal and Lignite Mines in the U.S. said. Article VI Coal and Rock Dust, Section 1a-Coal stated that dust should not be permitted to accumulate on haulage roads or on roadways of working places. Furthermore, Sections 2a and 2b stated that rock dust should be applied within 80 ft. of the faces in all open, unsealed rooms, haulage entries and parallel entries connected by open crosscuts. Sections 2a and 2b also stated that back entries should be rock-dusted for at least 1,000 ft. within the junction with the first active entry (Bureau of Mines, 1947).

Many employees in management violated the Federal Mine Code for Bituminous Coal and Lignite Mines by not adequately rock dusting. Mine inspectors Scanlan and Prez made these violations known to the mining company and their agencies. Many actions were taken by many parties, but few actions taken by management to abate hazards.

With the standard known, the next step is to understand the risk of an event resulting from the hazard. This information was also known. The U.S. Bureau of Mines conducted a test which indicated that coal dust required a presence of 33% incombustible matter to prevent ignition when no gas was present and 59% incombustible matter with gas present. This reference is to methane gas. These test results indicated that rock dusting with 33% incombustible matter can prevent explosions.

The investigation board came to this same conclusion. Based on these test results, the investigation team concluded that much of the untreated dust in the face regions was capable of initiating and propagating an explosion. The presence of coal dust in the treated areas was on the borderline safe side and was still effective (Bureau of Mines, 1947). The investigation team recommended that rock dust be applied up to and including the last open crosscuts in rooms and entries. The team also recommended that the face areas from the end of the rock dusted zone to the faces should be kept damp with water or a wetting solution.

Despite this information, many mine personnel at the time still considered methane gas a key ingredient in any mine explosion. In the case of Centralia Mine Number 5, no evidence showed methane gas as a cause or contributing factor in the explosion. The investigation team addressed this very issue and included in the mine report that "the outstanding lesson to be learned from this disaster is that mines, which liberate little or no methane gas, are not immune from widespread and tragic explosions if dry and dusty conditions exist therein and adequate measures are not taken to control the dust hazard" (Bureau of Mines, 1947). Perhaps more importantly, the investigation team went on to say that "if explosions of this type are to be prevented, it will be necessary to regard dry and dusty conditions in mines as being imminently dangerous in the future and to withdraw the men from the mine or portion thereof where such dangerous conditions exist, until appropriate measures have been taken to remedy such conditions" (Bureau of Mines, 1947).

Conclusion

Public sector safety professionals must be relevant, results-oriented, and purpose driven. A public-sector safety professional can be relevant to the needs of the public by focusing on results, not on activity. The story of the Centralia Mine Number 5 explosion is one of failure to focus on results. Because of that, the mine explosion killed 111 men. This essay described a classic focus on activity or doing things and no focus on achieving results. Great control was taken in inspections, writing letters, having meetings and visiting the mine; however, as the case demonstrated, activity does not create results.

Bibliography

Bureau of Mines Final report of mine explosion No. 5 Mine Centralia Coal Company, Centralia, Marion County, Illinois, March 25, 1947. Washington, DC.

CDC NIOSH. (December 28, 2001). Mining disasters. Washington, DC: Author, Centers for Disease Control and Prevention. Retrieved October 24, 2003 from http://www.cdc.gov/niosh/mining/data.

Illinois Labor History Society. Centralia Mine disaster. Retrieved July 20, 2005, from http://www.kentlaw.edu/ ilhs/centrali.htm.

Illinois Secretary of State. Illinois State Archives, Record Group 245.000: Department of Mines and Minerals. Retrieved October 26, 2003, from http://www.sos.state.il.us/departments/archives/di/245 002.htm

Martin, J.B. (March 1948). The blast in Centralia No. 5. *Harper's Magazine,* 1-38.

Martin, J.B. (2000). The blast in Centralia No. 5: A mine disaster no one stopped. In R.J. Stillman, *Public administration concepts and cases* (7th ed.), pp. 31-45. New York: Houghton Mifflin.

Stillman, R.J. III (2000). *Public administration concepts and cases.* New York: Houghton Mifflin. U.S. Department of Interior. (1947).

U.S. Mine Rescue Association. Historical data on mine disasters in the United States. Retrieved October 27, 2003 from http://www.usmra.com/saxsewell/historical.htm.

References

Garvey, G. (1997). Public administration: *The profession and the practice— A case study approach.* New York: Bedford/ St Martin's.

MSHA. A pictorial walk through the 20th century: Mine rescuers. Washington, DC: Author. Retrieved October 24, 2003, from http://www.msha.gov/CENTURY/RESCUE.

Scanlan, D. (April 24, 1947). Statement of Driscoll O. Scanlan before legislative committee at Centralia, IL.

Note: This article was originally published in the Spring of 2007, Perspectives Newsletter, Volume 6 Number 3.

This article was then selected as the Public-Sector Practice Specialty newsletter of the year and was published in the Council on Practice and Standards "Best of the Best from the 2005/06 Newsletters, 2006 Edition."

This article was selected as the "Best of the Best Newsletter Article for the Council of Practice and Standards for 2006-2007".

Essay 16-The Last Flight of the Space Shuttle

Introduction

At Arlington National Cemetery stands a monument to those who died in the Shuttle Challenger accident. It reads *"In Grateful and Loving Tribute to the Brave Crew of the United States Space Shuttle Challenger, 28 January 1986."* We recently passed the 25th anniversary of this sad event and now is a good time to relook the explosion of the space shuttle Challenger. This essay explores the circumstances that surrounded the explosion that destroyed the space shuttle killing those on board and touches on any relationship that exists between this tragedy and that of a later event that occurred with the space shuttle Columbia.

Background

The National Aeronautical and Space Administration (NASA) is the public organization that implements a Space Transportation System. This program involves launching and recovering space shuttles taking loads into space. NASA used four contractors to build and manage the program. The contractors included Martin Marietta Denver Aerospace, Rockwell International Corporation Space Transportation Systems Division, Morton Thiokol Corporation, and Rocketdyne. These contractors had specific areas of the program they supported.

Accident Phases

On January 27, 1986, several meetings were held to address the effect of cold weather on the launch. After a meeting of engineers at Mortin Thiokol the manager of the Ignition System and Final Assembly Solid Rocket Motor Project phoned the director of the Solid Rocket Motor Project to express concerns about the launch. A teleconference was held with personnel from Marshall and Kennedy Space Centers and Mortin Thiokol home office in which personnel from Mortin Thiokol recommend delaying the launch until 12:00 pm on January 28, 1986. At a meeting, later that day Mortin Thiokol personnel provided charts to Marshal and Kennedy Space Centers in which Mortin Thiokol personnel made it clear that they did not support a launch until the temperature

67

was 53° Fahrenheit (F). Later that day a Vice President for Mortin Thiokol agreed to the launch. The manager of the Ignition System and Final Assembly Solid Rocket Motor Project refused to sign the flight recommendation. Personnel from the plant signed the document and faxed it to the NASA. Unfortunately, during the evening and early morning hours the ice became a bigger problem for the shuttle. The morning of the launch Rockwell got involved and at 9:00 am on January 28, 1986 gave a confusing recommendation about launching. The flight was cleared to launch (Garvey, 1997).

The space shuttle launched at 11:38 am on January 28, 1986 from the Kennedy Space Center in Florida. The shuttle engines raised it above the tower and upward. All appeared to go well until Mission Control lost contact with the shuttle 73 seconds into the flight. Outside spectators could see a brilliant explosion that was the space shuttle. The explosion destroyed the shuttle and her crew. The external fuel tank was also destroyed. The two rocket boosters were blown clear and destroyed in flight.

On February 3, 1986, the president of the United States established an Executive Order 12546 directing a commission to investigate the explosion. Chapter four of the commission's report described the cause as "the consensus of the Commission and participating investigative agencies is that the loss of the Space Shuttle Challenger was caused by a failure in the joint between the two lower segments of the right Solid Rocket Motor. The specific failure was the destruction of the seals that are intended to prevent hot gases from leaking through the joint during the propellant burn of the rocket motor. The evidence assembled by the Commission indicated that no other element of the Space Shuttle system contributed to this failure" (Report, 2002).

Perhaps more importantly was the information contained in chapter 5 of the report in which the commission noted a contributing cause to the accident was that the decision to launch the Challenger was flawed and that those who made that decision were unaware of the recent history of problems concerning the O-rings and the joint. Furthermore, it noted that those who made the launch decision were unaware of the initial written recommendation of the contractor advising against launching at temperatures below 53° F and continuing opposition of the engineers at Thiokol after the management reversed its position. They did not have a clear understanding of Rockwell's concern that it was not safe to launch because of ice on the pad. Chapter 5 also noted that if decision makers knew all the facts, it is unlikely that they would have decided to launch flight 51-L on January 28, 1986 (Report, 2002).

Conflicts that need to be resolved

As the information became clear through the investigation it seems that NASA delegated the responsibility of the Space Transportation System program to too many. Furthermore, NASA used four contractors to build and manage the space shuttles. This caused a need for numerous levels of communications. One of the contractors was even selected based on low bid. Furthermore, the congressional interest in which state the work was done increased the stress on the agency.

There was a systemic weakness in the maintenance and evaluation program that allowed problems with the seals to go unchecked. This allowed the seal in the joint between the two lower segments of the right Solid Rocket Motor to fail causing the Challenger to explode. Furthermore, the information management system did not provide adequate information to those who made that decision to launch. This same system was side stepped by those who provided inadequate information to NASA managers.

Chapter 9 of the commission's report urged NASA to provide the President with a progress report on implementing the commission's recommendations. Five of those recommendations are very interesting:

- 2 - The Shuttle Program Structure should be reviewed.
- 3 - NASA and the primary Shuttle contractors should review all Criticality 1, 1R, 2, and 2R items and hazard analyses.
- 4 - NASA should establish an Office of Safety, Reliability and Quality Assurance to be headed by an Associate administrator, reporting directly to the NASA Administrator.
- 5 - Marshall Space Flight Center project managers stop managing in isolation and provide full and timely information bearing on the safety of flights.
- 9- NASA should establish a system of analyzing and reporting performance trends of such items (Report, 2002).

Also at Arlington National Cemetery is a monument to those who lost their lives in the Shuttle Columbia accident. It reads *"In Memory of the Crew of the United States Space Shuttle Columbia, 1 February 2003."* The Columbia Investigation Report has an interesting comment in the

Reader's Guide (National, 2003). The authors note "history reveals NASA has repeatedly demonstrated a lack of regard for outside studies and their findings" (page 10). As you read further in this report you find that there was not a genuine effort to comply with the Rogers Report from the Challenger accident. As Mr. Sean O'Keefe (National, 2003) accepted the Columbia Accident Investigation Report from the accident board president he noted "As an important step to change the culture of the agency, we have created the NASA Engineering Safety Center (NESC) at the agency's Langley Research Center in Hampton, Virginia, to provide comprehensive examination of all NASA programs and projects. The NESC, he said, could provide a central location to coordinate and conduct robust engineering and safety assessments across the entire agency. He went on to say that NESC was to play a key role in ensuring we return to flight safely and sustain a high level of engineering and safety excellence for every NASA program." This was a recommendation of the Challenger report years earlier to prevent accidents (National, 2003).

Conclusions

The monuments at Arlington are beautiful legacies of the sacrifice that crew members of the shuttles Challenger and Columbia made, but at what cost. This essay included the circumstances that surrounded the explosion that destroyed the space shuttle Challenger killing those on board. In the end, it seems, very little was done about the causes of this accident as proven in the findings of the explosion of the space shuttle Columbia. History will continue to repeat itself until hazards are abated rather than just included in reports.

Bibliography

Columbia Accident Investigation Report (2003). Retrieved on August 6, 2007 from http://www.nasa.gov/columbia/home/index.html

Garvey, Gerald (1997). Public Administration the Profession and the Practice a Case Study Approach, New York: Bedford/St Martin's.

National Aeronautics and Space Administration News Release 03-276 (2003). Retrieved on August 6, 2007 from http://www.nasa.gov/home/hqnews/2003/aug/HQ_03276_AOK_Acpt_CAIB.html

Report of the Presidential Commission on the Space Shuttle Challenger Accident (2002). Retrieved on August 6, 2007 from http://science.ksc.nasa.gov/shuttle/missions/51-l/docs/rogers-commission/table-of-contents.html.

References

Stillman, Richard J. III (2000). Public Administration Concepts and Cases, New York: Houghton Mifflin.

Washington Post Article "Summary: A Glimpse of Air Safety Survey (2007). Retrieved on January 2, 2008 from http://www.washingtonpost.com/wp-dyn/content/article/2007/12/31/AR2007123101462.html.

Note: This article was originally published in the Perspectives Newsletter, Volume 7 Number 3 in 2008 by the American Society of Safety Engineers, Council on Practices and Standards.

Essay 17-Safety Program and Project Evaluation

Introduction

To gain support for a program or project Safety, Health, and Environmental (SHE) professionals must use specific processes that allow management to understand the who, what, when, where, why, and how of the project or program. In most organizations, it is also important to conduct an evaluation of the results of the program or project. This essay will address the standard format of an evaluation that can assist SHE professionals to identify specific results from their efforts that will allow them to gain credibility for future endeavors.

To begin with there must be a need or reason to do the evaluation. There are a few questions that need to be answered to determine what will be evaluated and the purpose of the evaluation. The SHE professional then determines what information is needed to answer the questions; when the evaluation is needed and what resources are needed to conduct the evaluation. The next step is to determine how best to collect the information to evaluate. This information is used to develop the scope of the evaluation. The SHE professional must also determine who will use the evaluation and what questions the evaluation will seek to answer.

The key to collecting information is to first identify and gather correct information without gathering incorrect information. The SHE professional must determine what sources of information will be used. He or she will determine existing information, people who may be involved, and what needs to be observed. He or she then chooses data collection methods to use, which include surveys, case studies, and document reviews. The SHE professional then needs to determine what data collection procedures to use and when the data will be collected. This is followed by determining if a sample will be used as opposed to the population of data.

Once the data has been obtained it must be analyzed, interpreted, and communicated before it does any good. The SHE professional must determine how the data will be analyzed. Next, he or she must determine how the data will be interpreted and by whom. The last thing that must be done is to determine how the evaluation will be communicated and shared with those who need it.

This can be a long and difficult process unless it is done properly. As with any effort one must manage the process and this costs resources. The resources must be identified and obligated. Deliverables of the process that are important include the implementation plan (timeline and responsibilities), management or organizational chart (so the organization can be visualized), and the budget with sources of income and expenditures.

Example Program or Project Evaluation

To make this whole process easier to understand an example of an evaluation was developed that will be conducted to determine the effectiveness of the Missouri Motorcycle Safety Program (MMSP). Although the MMSP, Safety Council of the Ozarks, and Motorcycle Safety Foundation are real this example is not. It was done solely for this article; however, it does allow the reader to better understand how to conduct an evaluation by seeing one described. Author's notes are placed prior to each section to highlight important information.

Authors Note: The first step is to determine the focus of the evaluation. Sections 1-7 below describe how this might be done.

Focusing an evaluation

1. What will be evaluated?

The ability of the MMSP to train motorcycle riders to prevent accidents.

2. What is the purpose of the evaluation?

To measure whether the training conducted by the MMSP has reduced the number of motorcycle accidents in Missouri so that future decisions on continuing the program, funding levels to provide, and whether to use the current training materials can be made.

3. Who will use the evaluation?

Who/Users	How will they use the information?
Missouri State Legislators	To determine whether program is needed and funding levels.
Missouri Department of Revenue	To determine whether to continue to waive road test for motorcycle license applicants who have completed the Basic Rider Course [SM].
MMSP Administrator	To determine what training to provide.
Motorcycle Safety Foundation	To determine changes necessary to make courses more effective.
Training Site Sponsors.	To determine whether to continue to provide training.
Insurance Companies	To determine whether to continue to provide a discount to insured riders who have completed course.
Individual Motorcycle Riders	To decide whether to take the course of instruction or not.

4. What question will the evaluation seek to answer?

Is the training being conducted by the MMSP reducing the number of accidents experienced by its graduates?

5. What information is needed to answer the questions?

What I wish to know.	Indicators – How I will know it?
How many motorcycle riders in Missouri.	Query state driver's license bureau records and put information in a database by name.
How many licensed motorcycle riders in Missouri own a motorcycle?	Query state vehicle registration records and put the information in a database by name.
How many motorcycle riders had accidents in Missouri?	Check police and highway patrol records for accidents and put the information in a database by name.
Who completed a state sponsored motorcycle rider course from 2001 – 2006.	Query records from sponsored sites and put the information in a database by name.
How many of those who completed the course had accidents.	Analyze information from database.
Was the number of motorcycle riders who completed the course lower or higher than the number who are licensed and own a motorcycle?	Print the information from the analysis in a report.

6. When is the evaluation needed?

The results of this evaluation must be ready for presentation no later than June 30, 2008. This date will allow decision makers to have the information in time for 2010 decisions.

7. What resources are needed?

a. Time available to work on evaluation: 1 year (500 work hours) before the due date of June 30, 2008.

b. Money: $85,862.

c. People to assist in the evaluation – professional, paraprofessional, volunteers, and participants: 1 auditor, 1 statistician, 1 motorcycle instructor.

Authors Note: After completing the first section the SHE professional moves on to the second section answering the questions as he or she moves through section 8-13.

Collecting the information

8. What sources of information will be used?

a. Existing information: Accident records from state, counties, and cities. Training sites maintain training records for course attendees. The state maintains driving records and motorcycle registrations. Counties, cities, and state police districts have accident records.

b. People: No interviews or questionnaires are planned for this evaluation.

c. Observation: No observations are planned for this evaluation.

d. Pictorial records: No pictures are planned for this evaluation.

9. What data collection methods will be used?

a. Survey: No surveys are planned for this evaluation.

b. Case Study: No case studies will be reviewed for this evaluation.

c. Document review: The documents from the training sites showing the personnel who have completed the course; state driver's license records; state vehicle registration records; and local, county and state accidents investigation records. Instrumentation: A computer database program will be used to collect and analyze the information.

10. What data collection procedures will be used? When will the data for each method be collected? This is a process assessment for an on-going program so there will be no before or after data. Data during the time of the program will be used.

Method	Before Program	During Program	Immediately After	Later
Driver's License Review		January 1-30, 2008		
Accident Report Review		February 1-15, 2008		
Training Records Review		February 15-28, 2008		

Will a sample be used: Yes, but this will be a quasi-experimental evaluation. A sample of 15% will be drawn from both the control group (personnel not attending the course) and the group of those completing the course. The selection will be random. The statistician will format the database. The auditor will analyze and synthesize the material. The motorcycle instructor will act as subject matter expert and as a liaison with training sites.

11. How will the data be analyzed?

The accident experience of motorcycle riders who did not attend the training will be compared to the accident experience of motorcycle riders who did attend training. The comparison should show whether riders who completed the training have fewer accidents or not. The occurrence of fewer accidents could mean that the training provides the knowledge, skills, and abilities for a motorcycle rider to avoid accidents. This would mean that the course is effective and is meeting its objective.

12. How will the information be interpreted – by whom?

The information will be interpreted to verify or deny the thesis that the use of motorcycle training will reduce accidents. The auditor will develop the summary and presentation. A white paper will be developed and a short 15-20-minute presentation provided to officials.

13. How will the evaluation be communicated and shared?

To Whom	When/Where/How to Present
State MMSP Director	A presentation will be provided followed by questions and answers. The presentation date will be July 7, 2008. A hard copy of the white paper will be provided.
Motorcycle Safety Foundation Director	A hard copy of the white paper will be provided by July 7, 2008.
State Legislator Committee	A presentation will be provided to interested State Legislators the second week of July 2008. A hard copy white paper will be provided to each legislator.

Authors Note: Next the SHE professional must determine how he or she will manage the evaluation and obtain resources to complete the evaluation.

Managing the evaluation

14. Implementation plan: timelines and responsibilities.

Management Chart: The evaluation will be run by The Safety Council of the Ozarks in Springfield Missouri. They will pay for the evaluation, contract for services, and buy supplies. They will hire personnel needed to conduct the work. They will be reimbursed on a contract basis for office space and rent of furniture and equipment.

Budget: Sources of Income include requests for grants in the amount of $88,000.00 solicited from:

National Highway Traffic Safety Administration	$20,000
American Society of Safety Engineers	$12,000
American Motorcycle Association	$10,000
American Motorcycle Manufacturers Association	$20,000
Schnell Foundation	$8,000
Harley Davidson	$4,000
Victory	$4,000
American Medical Association	$10,000

$85,862.00. That amount includes salaries at $44,000, contracts and services at $41,393, and supplies and equipment $469.

Conclusion

The intent of this essay was to familiarize the SHE professional with the process and format of an evaluation that will help them speak the language of management and obtain permission and resources to carry out SHE programs and projects. This essay put forth a specific format as well as an example to better explain that format. If used, this method can enable the SHE professional to be acknowledged as a team player in management. With that acknowledgement comes additional resources and respect.

Note: This article was originally published in the Perspectives Newsletter, Volume 7 Number 2 in 2008 by the American Society of Safety Engineers, Council on Practices and Standards.

Essay 18-Lightning Protection Survey

Introduction

Many organizations provide public services out of doors and are therefore susceptible to all forms of severe weather. These organizations must provide for the notification and protection of the public while using facilities under these conditions. One severe weather incident can be easily planned for. Lightning need not scare or endanger the public using services and facilities out of doors. The recommended method is to conduct a lightning protection survey of all outdoor facilities with the results including a recommended primary and secondary source of protection from being struck by lightning. This should be followed by the development of a corrective action plan to ensure lightning protection is provided for each area. This essay presents a case study of an actual lightning protection survey and corrective actions taken to protect users of outdoor facilities from lightning strikes at a military installation. The author hopes that others will consider this process and use it themselves to prevent needless injuries and deaths from lightning strikes.

First Steps

Members of the Safety Office of a military installation conducted a lightning protection survey of all outdoor training facilities, ranges, firing points, and rappel sites. Thousands of soldiers, marines, sailors, and airmen are trained at this installation every year and the death or serious injury of even one of them from a lightning strike is unacceptable. This team prioritized the protection of personnel based on existing facilities and their use. The goal was for each person to have access to a facility with a lightning protection system where feasible. The results recommended a primary and secondary source of protection from being struck by lightning. In addition, standard operating procedures, lesson plans, and unit briefings were updated.

To start the process, the team plotted the location of all outdoor training facilities, ranges, firing points, and rappel sites on an installation map. Next they purchased five years' worth of lightning strike data for the installation. This was provided on a map with flash density or the number of flashes per square kilometer per year (Global Atmospherics, 2006). The data also included flash peak current frequency with both negative and positive polarity and a flash time trend plot by year. The data was extremely helpful and the team could not have determined which outdoor training facilities, ranges, firing points, and rappel sites to fix first without knowing which had the most likelihood of being struck by lightning. An office member found out about purchasing this information from attending the annual conference of the Lightning Strike Electric Shock Survivors International, Inc. (Lightning Strike, 2009)

The team then developed a survey instrument to identify and collect specific data on all outdoor training facilities, ranges, firing points, and rappel sites. This instrument included consideration of the following:

- Ranges and fixed training areas (included how many people trained on site at one time and how often these sites were used).
- Identified existing facilities and indicated where no facilities were available.
- Identified evidence of a lightning protection/grounding system in place. Annotated which system was installed: grounding system or a lightning protection system.
- Noted what protocols and/or standard operating procedures were in place to notify persons at risk from the lightning threat.
- Identified secondary areas for dispersion of people in the event of a lightning storm if lightning protection shelters were not available or people were not near a shelter.

Facilities were prioritized based on the following:

- Lightning protection system in place.
- Grounding system in place.
- Ungrounded facility.
- Open area.

There were some assumptions the team made that included that no place is safe from the lightning threat; however, some places are safer than others. Large enclosed structures and substantially constructed buildings tend to be much safer than smaller or open structures. The risk for lightning injury depends on whether the structure incorporates lightning protection, construction materials used, and the size of the structure (NFPA, 780).

Small open shelters are common on golf courses, athletic fields, parks, roadside picnic areas, schoolyards, as well as many other locations. Many of these shelters are built to protect against rain or sun, not lightning. Although there is no such thing as a lightning-proof small outdoor shelter, a properly designed and installed lightning protection system can significantly reduce the risk of a lightning strike.

Protection Standards

For this survey, a lightning protection system for an ordinary structure included air terminals, down conductors, and ground terminals. These three elements of the system formed a continuous conductive path for lightning current, with all connections between the elements typically accomplished by bolting or welding. The function of such a system was to intercept lightning and safely direct the current to the ground. For structures with metal roofs with a thickness of 3/16 inch or greater, the roof played the role of the air terminals. The structural metal framework to include the metal support posts played the role of down conductors because they were electrically continuous. Sometimes the ground terminal was a buried bare conductor wire encircling the structure known as a loop conductor. Such grounding was used to intercept ground surface or underground electrical arcs that may develop and move toward the structure from a nearby object being struck by lightning. The closer the structure approached a Faraday Cage, the better its interior was protected from lightning effects. In the absence of the three-element lightning protection system, the structure was considered unprotected for lightning.

Small shelters without lightning protection were to be avoided during thunderstorms, particularly if they were located at a high elevation or near a tree or a small group of trees dominating the area. If there was no better choice, shelters in relatively low areas were to be used, preferably surrounded by many trees of approximately the same height.

A small shelter equipped with a properly designed and installed lightning protection system provided reasonable protection from lightning. It was essential; however, that a person inside the shelter not touch any element of the lightning protection system and positioned him or herself at approximately the same distance from all down conductors. A small shelter, even one protected as described here, was viewed as the last resort option. Better-protected shelters such as large buildings were sought when possible.

A properly protected shelter had to have at least one air terminal, at least two down conductors on two diagonally opposite sides of the structure, and ground terminals connected to the down conductors. Two designs for ground terminals in common soils were used.

- Vertical ground rods; at least one for each down conductor, interconnected by a loop conductor driven into the ground.
- Ground terminals included horizontal conductors, at least one for each down conductor, buried and extended away from the shelter beyond the roof drip line. This design also included a loop conductor.

In both designs, the addition of a buried metal mesh within the shelter perimeter, connected to the ground terminals, was a preferred option. Another option was for the floor to be made of asphalt; rock or wood to further reduce the lightning hazard for people inside the shelter. As an alternative lightning protection systems consisting of grounded overhead wires suspended above the shelter on separate poles were deemed acceptable. The loop conductor mentioned above could be employed here too.

Rod-type air terminals were solid or tubular copper rods. Air terminals were placed on ridges of pitched roofs and around the perimeter of flat or gently sloping roofs. Down conductors were stranded cables. The bends of conductors were 90 degrees and the radius of the bends was more than 8 inches. Down conductors were covered with insulating material resistant to impact and climate conditions. Vertical ground rods were made of solid copper, and horizontal conductors were made of stranded copper cables. Aluminum conductors were not used.

Next Steps

Once the survey was done for all outdoor training facilities, ranges, firing points, and rappel sites, those that had proper protection had signs posted on them to show which facility was to be used for protection in the event of a lightning storm. A corrective action plan was also developed to identify those sites that did not have appropriate lightning protection. The corrective action plan was designed to make all corrections over a five-year period. Unfortunately, there were some low-risk improvements that would take longer than five years. These improvements were made as funds became available, but after higher risk improvements were completed. The total cost to provide facilities with a lightning protection system where feasible was $11,080,781.00. If you broke that down by year the costs were:

- 1st year $117,435.00
- 2nd year $897,412.00
- 3rd year $2,330,464.00
- 4th year $1,274,578.00
- 5th year $1,059,656.00

The cost for structures listed as priority 6 (as funds are available) is $5,401,236.00.

There were some facilities that provided minimum protection. In a few cases, the team identified the correct procedures to be used to prevent being struck by lightning even in open areas.

Within 120 days of completion of the lightning survey, the team installed a sign on each facility identified as a "Lightning Protection Shelter" so they were easy to recognize. They also developed a training package to train personnel about the hazards of lightning and ways they could protect themselves. The bleachers inside windscreens that were identified as "Lightning Protection Shelters" were grounded to the shelter.

Summary

This essay provided the reader with a case study of the process used by a military installation to identify the need for and plan to have at least one lightning protection shelter for every outdoor training facility, range, firing point, and rappel site. This same methodology works for golf courses, athletic fields, parks, roadside picnic areas, schoolyards, camping grounds, swimming beaches, and elsewhere in the public sector. This case study also showed just how expensive it can be to provide lightning protection for users of outdoor facilities. Before the cost can even be known, the SHE professional must work the process to determine exactly what the hazard is and how to control it. Only then can the cost be known and the work of locating resources begins. The key is that this cannot be done overnight or all at once; however, step by step it can be done.

Bibliography

Global Atmospherics, Fort Leonard Wood Facility Site Analysis, January 1, 1997 to December 31, 1002, December 2006.

NFPA 780, Standard for the Installation of Lightning Protection Systems (1997), National Fire Protection Association, Quincy MA.

References

Installation Requirements for Lightning Protection Systems - UL96A (1998), Underwriters Laboratories, Northbrook IL.

Lightning Strike Electronic Shock Survivors International, Inc. Retrieved on August 24, 2009 from http://www.lightning-strike.org/DesktopDefault.aspx.

National Weather Service, web page. Retrieved on August 24, 2009 from http://www.lightningsafety.noaa.gov/.

NFPA 70 (1999), National Electrical Code, National Fire Protection Association (NFPA), Quincy MA.

US Army Training and Doctrine Command Memorandum, Subject: Lightning Protection for Personnel, February 22, 2002.

*Note: This article was originally published in **The Perspectives Newsletter**, Volume 9, Number 2 in the Winter of 2010 by the American Society of Safety Engineers, Council on Practice and Standards.*

Essay 19-Influenza Pandemic Preparation Redux

Introduction

In 1917, a flu virus infected the residents of rural Haskell County, Kansas. Later that year young men from Haskell County went to Camp Funston, Kansas to receive Army training in preparation for combat duty in World War I. The soldiers at Camp Funston slept and worked in close quarters through a very cold winter without adequate heat (Barry, 2004). The author of this essay believes the flu pandemic that killed approximately 40 million people worldwide began in Haskell County, Kansas.

There is a new strain of influenza that is estimated to cause 40% absenteeism in the workplace. This virus is believed to can kill 50% of its victims and can turn the human immune system against young and healthy adults. This virus is unlike any since the great influenza pandemic of 1918 (Department of Commerce, 2006).

Influenza is a normal part of life in the U.S. and people are exposed to influenza every year. Of those that contract the virus, approximately 35,000 will die. The elderly, young children, and people with immune system weaknesses are more susceptible to the virus. A vaccine is developed each year based on the strain of virus infecting humans. Many people make a choice not to receive an influenza vaccination and never get the flu; however, medical experts recommend that the elderly, young children, and people with immune system weaknesses get a flu shot every year. This essay was written to prepare Safety, Health, and Environmental (SHE) professionals to be called upon to help local and state governments prepare for and react to an influenza pandemic when and if a pandemic occurs. This essay addresses occupational health recommendations that if implemented will reduce the potential for a pandemic to spread.

What is Influenza?

Influenza is a virus that normally affects birds. The virus in birds may be contracted by human beings through contact with the body fluids of an infected bird. When contracted in this way, the human suffers the effects of the strain of virus but cannot pass it on to another human being.

If this bird virus infects a swine or human who is infected with a human form of influenza, the two forms may join to form a third virus capable of being transmitted human-to-human. In his book the Great Influenza, John Barry (2004) describes it as "The influenza virus not only mutates rapidly, but it also has a "segmented" genome. This means that its genes do not lie along a continuous strand of nucleic acid, as do genes in most organisms, including most other viruses. Instead, influenza genes are carried in unconnected strands of RNA. Therefore, if two different influenza viruses infect the same cell, "assortment" of their genes becomes very possible."

Influenza usually develops in Asia and moves around the world. In the U.S., health officials monitor the strains. Vaccinations are developed for each strain and usually provided to the U.S. population prior to the arrival of that strain of influenza. People who are vaccinated for one strain are not protected from any other strain moving through the population.

What is a Pandemic Flu?

The U.S. Department of Health and Human Services (Pandemic Flu 2006) web site "PandemicFlu.gov" provides essential information. This site defines Pandemic flu as "A virulent human flu that causes a global outbreak, or pandemic, of serious illness."

The White House fact sheet, Safeguarding America against Pandemic Influenza (Fact Sheet, 2006), states that "Pandemic influenza poses a greater danger than seasonal flu. Most Americans are familiar with influenza or the "flu" as a respiratory illness that makes hundreds of thousands of people sick every year. For most healthy people, the flu is not usually life-threatening. Pandemic influenza is another matter. It occurs when a new strain of influenza emerges that can be transmitted easily from human-to-human and for which people have no immunity. Unlike seasonal flu, it can kill the young and healthy as well as the frail and sick."

Bird Flu

Most discussions about Pandemic Influenza usually include the H5N1 virus because it has proven to be especially tenacious. The virus has infected large populations of birds in Indonesia, Vietnam, Cambodia, and China, among other countries. The infected birds were destroyed to prevent the spread of the virus. Human deaths from H5N1 have occurred in Azerbaijan, Cambodia, China, and Indonesia, among other countries. For the actual number of cases as of March 24, 2006, see Table 1 (World Health Organization, 2006).

Country	2004		2005		2006		Total	
	cases	deaths	cases	deaths	cases	deaths	cases	deaths
Azerbaijan	0	0	0	0	7	5	7	5
Cambodia	0	0	4	4	1	1	5	5
China	0	0	8	5	8	6	16	11
Indonesia	0	0	17	11	12	11	29	22
Iraq	0	0	0	0	2	2	2	2
Thailand	17	12	5	2	0	0	22	14
Turkey	0	0	0	0	12	4	12	4
Viet Nam	29	20	61	19	0	0	93	42
Total	46	32	95	41	42	29	186	105

Table 1 - Cumulative number of confirmed human cases of Avian Influenza A/ (H5N1) reported to World Health Organization

The H5N1 may not be the actual strain of virus that causes the pandemic; however, if it or one like it does cause a pandemic, the death toll could be staggering. This caused the U.S. Government to take steps to prepare for and respond to a pandemic. The White House published a National Strategy for Pandemic Influenza in November 2005. This document provided a high-level overview of the approach that the U.S. Government would take to prepare for and respond to a pandemic. The U.S. strategy planned for state and local governments to prepare themselves and their communities.

The goal of this strategy was to "Enable the US Government to collaborate fully with international partners to attempt containment of a potential pandemic whenever sustained and efficient human-to-human transmission was documented, and make every reasonable effort to delay the introduction of a pandemic virus to the U. S. If these efforts fail, responding effectively to an uncontained pandemic domestically will require the full participation of all levels of government and all segments of society" (National Strategy, 2005).

Identify Essential Functions and Services

All levels of government have a responsibility to ensure continuation of essential functions and services for its citizens. Law enforcement and firefighting are just two examples of essential functions. Once the essential functions and services have been identified, officials must identify supporting activities and capabilities. Supervision and control, facility security, communications, personnel accountability and recall, facility management, legal support, procurement, records protection, and computer and IT are supporting activities. Communications, phone service, teleconferencing, secure networking, internet access, business supplies, and trained personnel are capabilities.

Protecting the Health of Employees

Once the essential functions and services have been identified, the SHE professional can begin to protect the health of employees who come to work to provide those functions and services. The first rule is that any employee not needed to perform or support an essential function or service should work from an alternate location to prevent the spread of the virus.

Managers and supervisors need to prevent employees from contracting the flu virus while at work. They must prevent exposure by limiting social contact between employees while at work and raising awareness of personal hygiene to promote overall health.

SHE Professionals

SHE professionals should monitor pandemic information from the Department of Health and Human Services (HHS), the World Health Organization (WHO), and the Department of Labor (DOL), and distribute this information.

In addition, they should coordinate with the organization's health office or local health facilities to support ill or injured employees. These individuals should also coordinate for vaccination of personnel when vaccines become available.

SHE professionals should coordinate with security and facility personnel to provide medical personnel at building entrances for "report to work" evaluations. "Report to work" evaluations consist of a health care professional who will view occupants as they enter the building and identify those that show outward symptoms of influenza. Any individual showing outward symptoms of the virus should be refused access.

Managers and Supervisors

Managers and supervisors are in the best position to enact policies and procedures to limit exposure and the likelihood that employees will contract influenza at work. They should act based upon guidance and policy issued by HHS and DOL, including following HHS recommendations for personal protective equipment (e.g., surgical/procedure masks and gloves) for essential employees who must report to work.

To make personal protective equipment viable, organizations should stockpile the personal protective equipment recommended by HHS at a ratio of 120% percent of the employees necessary for essential functions and services for approximately 15 days. This will allow for employee and visitor replacements as needed. Managers and supervisors must order the equipment and notify employees of the equipment to be issued, how the employees will obtain it, and how the employees will be trained.

Tools for Prevention

Only bring the employees to work that need to be there. To accomplish this, one can use the concept of "reasonable risk." Reasonable risk is the trade-off of the risk of having employees report to work and possibly getting the flu versus the risk of not completing non-essential functions and services.

Use the concept of social distancing to limit exposure for employees that must come to work. Social distancing is keeping employees at least three feet from each other. There are several methods that may be used to create social distancing. Table 2 outlines the more important ones.

When possible, let employees work from home. To facilitate this, managers and supervisors must determine how many personnel will need access and the type of access, e.g., e-mail or Virtual Private Network (VPN) prior to the onset of a pandemic.

Get the right medical resources for the organization. Use existing medical resources, implement agreements to staff health offices, and identify additional employee assistance resources that may be needed.

Maintain contact with HHS to learn about the availability and distribution of anti-viral agents and pre-pandemic and pandemic vaccines, to ensure designated employees receive them.

Summary

The H5N1 virus served as a wake-up call to public officials about the dangers that an influenza pandemic might pose. If governments are to prepare for a pandemic, they must be able to provide essential functions and services while experiencing absenteeism rates up to 40% and death rates up to 50% for employees contracting the virus.

Work on the 1918 Influenza Pandemic continues. Dr. Johan Hultin worked for over 50 years to obtain tissue samples from victims of the 1918 pandemic buried in the permafrost of Alaska (Kleffman 2006). Dr. Jeffery Taubenberger credits Dr. Hultin's work for enabling his team of experts to identify the entire genetic code of the virus. Dr. Taubenberger has even brought it back to life to help fight the Avian Influenza (Kleffman 2006). Perhaps the greatest work is yet to come. If local and state governments are to provide essential functions and services during a pandemic, they must identify and apply methods to prevent and control the virus now.

Bibliography

Barry, John, The Great Influenza, The Epic Story of the Deadliest Plague in History, Penguin Books, 2004.

Department of Commerce Department Plan for Pandemic Influenza, U.S Department of Commerce, March 24, 2006.

Fact Sheet: Safeguarding America against Pandemic Influenza, White House. Retrieved on March 30, 2006 from http://www.whitehouse.gov/infocus/pandemicflu/.

Kleffman, Sandy. Researcher's quest to revive virus spanned decades. The Free Lance Star, Fredericksburg, Virginia, March 26, 2006, page A8.

Kleffman, Sandy. U.S. Scientists bring 1918's deadly influenza virus back to life, The Free Lance Star, Fredericksburg, Virginia. March 26, 2006, page A-8.

National Strategy for Pandemic Influenza, Office of the President, Homeland Security Council, November 2005. Retrieved on March 30, 2006 from http://www.whitehouse.gov/homeland/pandemic-influenza.html.

Pandemic Flu, Department of Health and Human Services. Retrieved on March 30, 2006 from http://www.pandemicflu.gov/.

World Health Organization (WHO), Epidemic and Pandemic Alert and Response (EPR), Cumulative Number of Confirmed Human Cases of Avian Influenza A/ (H5N1) Reported to WHO. Retrieved on March 30, 2006 from http://www.who.int/csr/disease/avian_influenza/country/cases_table_2006_03_24/en/index.html.

Essay 20-Inter-Agency Forum on Climate Change

Introduction

The author attended the April meeting of the Interagency Forum on Climate Change Impact and Adaptations in Washington, DC. The meeting was sponsored by the National Aeronautics and Space Administration and was held in April 2009. This meeting was a half day of sessions focused on "presenting key concepts and issues related to the impacts of and potential adaptations to climate change" in the hopes that America can "manage the unavoidable while seeking to avoid the unmanageable" (NASA 2009).

The Interagency Forum

The forum is more correctly described as a Federal Government interagency working group. The purpose of this interagency working group is to provide an informal forum for sharing data, approaches, scenarios, and expertise relative to forecasting and responding to climate change impacts (Climate Change 2009). The brochure for the forum provided the web site for this forum, but when the author visited it, there was no material there. The actual address for the web site is provided in the reference section of this essay. The interagency forum meets every six to eight weeks in the Washington, DC area. Currently, employees from various Federal Agencies participate in the interagency forum.

The Forum April Meeting

The forum provided four hours of information. Attendees were also provided copies of the book, A Federal Leader's Guide to Climate Change. As the reader can see from the agenda at Table 1, the information covered everything from climate change 101 to advanced discussions about regionally specific issues. The author found the presentations very insightful and well done. The presenters were obviously very knowledgeable in their subject area and had a passion for the work.

93

Olga Dominguez with NASA and Patricia Rivers with the US Army Corps of Engineers provided opening statements about the issues of concern involved with climate change and some of the changes already being seen and experienced.

Steve Seidel with the Pew Center on Global Climate Change provided basic information about exactly what climate change adaptation was. This presentation looked at the basics of climate change, what can occur, and what is most likely to occur.

Interagency Forum on Climate Change Impacts and Adaptations	
Opening Remarks	Olga Dominguez, Office of Infrastructure, NASA Patricia Rivers, Military Programs Directorate, US Army Corps of Engineers
Climate Change 101: What is Climate Change Adaptation?	Steve Seidel, Pew Center on Global Climate Change
Adapting to Climate Change Challenges – Learning from the New York City Experience	Dr. Cynthia Rosenzweig, Goddard Institute for Space Studies, NASA
Adapting to Climate Change in Florida: Some Regional and Local Considerations	Dr. Duane E. De Freese, Science and Biology Department, University of Central Florida
Climate Change Impacts and Water Management	Dr. Kathleen D. White, PE, Cold Regions Research and Engineering Laboratory, US Army Corps of Engineers
Report on Federal Activities Related to Climate Change Adaptation	John Stephenson, Natural Resources and Environment, Government Accountability Office
A Federal Leader's Guide to Climate Change	Dr. Rachel Jonassen, LMI Research Institute

Table 1 - Interagency Climate Change Forum, April 21, 2009 Meeting Agenda

Dr. Cynthia Rosenzweig with NASA provided a solid presentation on the experiences that New York City underwent due to climate change. This was followed by Dr. Duane De Freese of the University of Florida telling attendees about the experiences Florida underwent due to climate change. These two presentations worked very well together and highlighted consistencies and differences in the way climate change is forcing cities and states to adapt.

Dr. Kathleen White with the Army Corps of Engineers next discussed the impacts that climate change had on the ability to manage water. This could be one of the more important adaptations Americans will have to make to ensure that fresh drinking water is available and flooding can be controlled.

Next, John Stephenson of the Government Accountability Office highlighted what the Federal Government did to adapt to climate change and the way it is affecting America. His presentation was followed by Dr. Rachel Jonassen with LMI Research Institute, in which she discussed the development and production of a valuable book titled "A Federal Leader's Guide to Climate Change." This book provides Federal Leaders with a great deal of information to guide them in the process of adapting to climate change and allowing the Federal Government to change programs, legislation, and processes before climate change forces adaptation.

Summary

The meeting was well worth the four hours it took to attend. The information presented at the forum was second to the information provided in the book. The author was impressed with the forum proceedings. The author also feels confident that any information requested from the forum would be high quality. If the reader is with a Federal Agency or a private sector agency that is doing research into climate change adaptation, this is a good source of information.

Reference

Climate Change Interagency Working Group Web Site. Retrieved on July 31, 2009 from http://www.fedcenter.gov/Articles/index.cfm?id=9915&pge_id=1606.

Understanding Climate Change Adaptation; Managing the
   Unavoidable While Seeking to Avoid the Unmanageable, brochure,
   National Aeronautics and Space Administration, April 21, 2009.

*Note: This article was originally published in The Envirometer
Newsletter, Volume 9, Number 1 in the Winter of 2010 by the
American Society of Safety Engineers, Council on Practice and
Standards.*

Essay 21-Continuity of Operations Planning

Challenge

It should be the policy of every public organization to have in place a comprehensive and effective program to ensure continuity of essential functions under any circumstances. As a baseline of preparedness for the full range of potential emergencies, all public-sector organizations should have in place a viable Continuity of Operations (COOP) capability which ensures the performance of their essential functions during any emergency or situation that may disrupt normal operations (Federal Protection Circular 65, 1999). Given such an overarching challenge, COOP planning is a must for public sector organizations to ensure they can provide needed services for the citizens who depend on them in an emergency. There are several objectives to any COOP plan, which include:

1. Test the alert, notification, and activation system.
2. Perform selected essential functions from an alternate site.
3. Access vital files and databases from an alternate site.
4. Communicate effectively from an alternate site.
5. Receive, process and analyze, and disseminate information.
6. Validate support systems for 24-hour operations.

Planning

A March 2008 article in the Government Computer News (National Continuity) stated that "COOP (Continuity of Operations) planning is the disciplined planning you do in advance to respond to a natural or man-made emergency. If your agency/office needs to relocate your COOP is your coordinated, efficient action to keep operating" (Telework, 2008). Some tasks that public sector organizations will have to provide to allow them to operate and assist the public include:

1. Providing for the safety and well-being of employees
2. Providing administrative and facilities management and support services
3. Providing travel and transportation services

4. Identifying all affected real and personal property
5. Providing mail and courier delivery services
6. Coordinating facility repair and operations
7. Acquiring space and facilities

This plan will allow a public organization to continue its work with little or no disruption in service; however, it takes a lot of work to do well. For example, at the national level, "continuity planning also requires coordination with state, local, tribal, and territorial governments as well as the private sector" (Telework, 2008). Planning is required with private sector organizations "because the private sector owns the vast majority of the nation's infrastructure, we have a nation that is a "system of systems" that is incredibly integrated" (Telework, 2008). This planning must also include information technology where "Networks must connect. Applications must be streamed. Computer screens must look familiar. Security must be maintained at all costs and Government business must continue" (Telework, 2008).

There is a lot of guidance from the Federal Government that can be used in the planning process. Federal Protection Circular Number 65 "includes plans and procedures that delineate essential functions; specify succession of office and an emergency delegation of authority; provide for the safekeeping of vital records and databases; identify alternate operating facilities; provide for interoperable communications and validate the capability through tests, training, and exercise" (Telework, 2008). This circular requires the following:

1. Identify succession for senior members of the organization
2. Delegate full authority in order of succession
3. Identify alternate facility or facilities for critical personnel
4. Demonstrate a general level of understanding of the COOP process among employees
5. Train employees as to their individual roles (if any) regardless of the organization's level of involvement during a COOP activity
6. Direct employees to involve themselves in and support COOP activities fully possible, to include training scenarios
7. Maintain vital records and databases in electronic form at a backup location to meet operational responsibilities following the activation of a COOP

There is also Federal Continuity Directive Number 1 (FCD#1), which "provides direction to the Federal Executive Branch for developing continuity plans and programs. What FCD#1 says is that it is just not good practice to plan to have continuity planning, it is mandatory practice" (Telework, 2008).

In addition, there is Federal Continuity Directive Number 2 (FCD#2), which "implements the requirements of FDC#1 and provides guidance and direction for identification of MEFs [*Mission Essential Functions*] and potential PMEFs [*Primary Mission Essential Functions*]. It includes guidance and checklists to assist departments and agencies in assessing their essential functions through a risk management process and in identifying potential PMEFs that support the National Essential Functions (NEFs)" (Telework, 2008).

Teleworking

Teleworking (employees working from an alternate site or home using telephone and computer) is one element of planning that must be considered. Teleworking can allow public sector employees to work from home or an alternate location when a public-sector facility is damaged, cannot be reached, or is in the path of an impending disaster. Standards and guidance must be in place prior to any emergency to ensure employees have the equipment necessary to perform their work. This means "To support the technology components critical for telework translates into spending precious dollars in areas such as web-based applications, BlackBerry devices, laptops, and remote email access, which allow for increased telework at low incremental cost" (Telework, 2008). Richard Walker also noted in Federal Computer Weekly that telework "means creating the capability for employees to work from home or other remote locations and having an information technology infrastructure that is robust enough to support remote access to vital agency computer systems" (OPM, 2008). In that same article, Walker noted that "The Nexus between COOP and telework has become increasingly important in recent years, underscored and reinforced by high-magnitude events such as the 2001 terrorist attacks and hurricane Katrina" (OPM, 2008).

The Office of Personnel Management is the proponent for personnel policy for the Federal Government. In their publication, Federal Manager's/Decision Makers Emergency Guide, key steps to facilitate telework include:

1.   Develop a cadre of regularly scheduled "core" teleworkers.

2. Permit teleworkers to experience working off site and learning to communicate electronically with colleagues and clients by doing it regularly.
3. Permit supervisors and managers to experience managing employees without face-to-face contact. (Federal Managers, 2008)

Security

There are always security concerns involved with moving an organization to an alternate site when responding to an emergency. Specifically, public administrators experience a great deal of concern over the security of the information technology equipment and transmissions. In his article for Federal Computer Week, Alan Joch tells us that "The right mix isn't purely technical. With the right selection of hardware and software, agencies can ensure that established security policies remain in effect during an emergency" (How Secure, 2008). Alan Joch recommends using a class of technology known as network access controllers to vet remote machines, two–factor authentications for access, data encryption technologies [for data encryption], and virtual private networks [for secure access] (How Secure, 2008).

In addition, security planners must also consider the physical security of any alternate site. It is essential that planners conduct risk assessments of possible alternate sites so that risks can be considered along with other criteria to determine the best fit. These risk assessments should identify the potential security weaknesses of a facility. Effort should then be expended to reduce some of the risks for selected facilities. A good example is that many public buildings now are using "stand-off" to reduce the impact of a blast on the building. A second example is the installation of blast windows that can withstand an explosion. Planning officials must make sure that any alternate facility includes "stand-off" and blast protective windows when selecting an alternate site.

Summary

This essay has touched on a variety of topics that all deal with public organizations putting in place a comprehensive and effective program to ensure continuity of essential functions under all circumstances. The topics of planning, telework, and security were each explored. To ensure that public organizations can respond in an emergency, it is essential that all aspects of the response be planned for. Without this planning, it will be very difficult if not impossible for a public organization to respond, and if the response is weak, the voters can and should bring new public servants into office to do a better job.

Bibliography:

COOP: The telework connection, by Richard W. Walker, Federal Computer Week, June 16, 2008.

Federal Manager's/Decision Maker's Emergency Guide, Office of Personnel Management, US Government. Retrieved on September 30, 2008 from https://www.opm.gov/emergency/PDF/ManagersGuide.pdf.

Federal Protection Circular 65, July 26, 1999. Retrieved on September 30, 2008 from http://www.fas.org/irp/offdocs/pdd/fpc-65.htm.

How Secure is your COOP, by Alan Joch, Federal Computer Week, June 23, 2008.

National Continuity, Government Computer News, March 31, 2008.

OPM's Best Practices for COOP, by Richard W. Walker, Federal Computer Week, June 16, 2008.

Telework Is Taking Off, Government Computer News, March 31, 2008.

Note: This article was originally published in the Perspectives Newsletter, Volume 8, Number 3 in 2009 by the American Society of Safety Engineers, Council on Practice and Standards.

Essay 22-Increase Results through Synergy

Introduction

Public Safety is a very broad subject. It involves the protection of the public from injury and illness from a variety of sources that include criminal as well as unintended acts or what are often called accidents. In many cases the victim is injured by themselves through commission or omission of an act. In today's world, public safety is becoming more visible and involves safety, environmental protection, crime prevention, fire prevention, and health promotion. Unfortunately, each of these areas is a stand-alone organization with little interface with the others and in some cases duplicating some of the same work. Furthermore, many of these offices are shorthanded and if they could save resources through any new approach those savings could be used to fill the shortage of resources. Just as important is the fact that the public does not get a uniform approach to prevention. This can cause many of them to not take the threats seriously. At appendix A is the original proposal for this research project. My thesis is that limited resources and overlapping programs provide the Army with weak and inconsistent safety messages followed by an inability to provide all the services necessary and if these organizations were combined or organized to pull together there could be one large prevention program focused in one direction for the good of all with the overlap eliminated and those resources used to fill resource shortfalls.

Public Safety in the U.S. Army

In the U.S. Army, public safety is the purview of at least five different organizations. The military police take care of physical security, crime prevention, and investigations. The fire department is responsible for fire prevention and protection. While the Safety Office is responsible for boating and water safety, accident prevention, motor vehicle accident prevention, fire prevention and injury, firework safety, and hot and cold weather injuries. The Occupational Health Clinic is responsible for hot and cold weather injuries, sunburn, stress, exercise safety, mold, and the effects of drug and alcohol. That is a lot of work to go ground.

However, despite the workload many of these offices are resourced at 65-85% of the requirements. This equates to between 15-35% of the work that the U.S. Army wants done to protect people and resources is not being done. This leaves a very large gap. Within that gap can be events that result in injuries, property damage, disabilities, and even death. The secret here appears to be using resources in a more focused way to identify the most important things to do, prevent any overlap, and coordinate the implementation of the program.

Safety and Occupational Health Management

Within the responsibilities of the Safety and Occupational Health Program are certain ones that directly relate to public safety as opposed to occupational safety. It is important to look at the structure of this program. Army Pamphlet 10-1 (1994) is a great resource to identify the offices responsible for the safety program. There is an Assistant Secretary of the Army for Installation, Logistics, and Environment. There is also a Deputy Assistant Secretary of the Army for Environmental Safety and Occupational Health. The Army also has a Director of Army Safety on the special staff of the Chief of Staff of the Army. The individual that holds this position is also the Commander of the U.S. Army Safety Center.

Army Regulation 385-10 (2000) spells out the Army Safety Program. In that regulation, the Office of the Assistant Secretary of the Army for Installations, Logistics, and Environment is the principal consultant to the Secretary of the Army for Army safety and occupational health matters. The Assistant Secretary of the Army for Installations, Logistics, and Environment is the Army's designated safety and occupational health official. See Table 1 for a list of his responsibilities. Army Regulation 385-10 goes on to say Commanders at all levels will be responsible for protecting personnel, equipment, and facilities under their command; effective implementation of safety and occupational health policies; and the integration of the risk management process into their safety and occupational health program.

- Approve policies, issue directives, make recommendations, and issue guidance on Army safety and occupational health plans, programs, and risk management integration.
- Initiate programs, actions, and taskings to ensure adherence to Department of the Army and Department of Defense safety and occupational health policies.
- Review and evaluate programs for carrying out approved safety and occupational health policies and standards.
- Serve on boards, committees, and other groups pertaining to safety and occupational health, and represent the Secretary of the Army on safety and occupational health matters outside the Department of the Army.
- Participate in the planning, programming, and budgeting of safety and occupational health.
- Serve as Functional Chief for the Safety Management Career Program

Table 1 - Duties of the Assistant Secretary of the Army for Installation, Logistics, and Environment

As with any program the commander cannot do it all. He or she hires a Safety Manager to develop and run a program at the installation to assist them in meeting their responsibilities outlined in Table 2.

- Designate a command safety and occupational health official to exercise staff supervision over safety and health, risk management, and accident prevention activities.
- Ensure that the designated command safety and occupational health official is a member of the commander's special staff reporting directly to the commander.
- Ensure the designated command safety and occupational health official meets Office of Personnel Management standards for the positions of occupational safety and health.
- Organize and staff a comprehensive safety office under the direction of a designated command safety and occupational health manager.

Table 2 - Requirements of Installation Commander for Safety.

Chapter 2-1 of Army Regulation 385-10 describes the structure of all safety offices. This office will organize and administer a safety program that includes the programs outlined in Table 3.

Accident reporting	Workplace safety	Transportation safety
Family and off-the-job safety	Range safety	Explosive safety
Aviation safety	Tactical safety	Radiation safety
System safety		

Table 3 - Duties of an Installation Safety Office.

This safety office is very seldom the same from one installation to another; however, I have tried in Figure 1 to lay out what a typical office might look like.

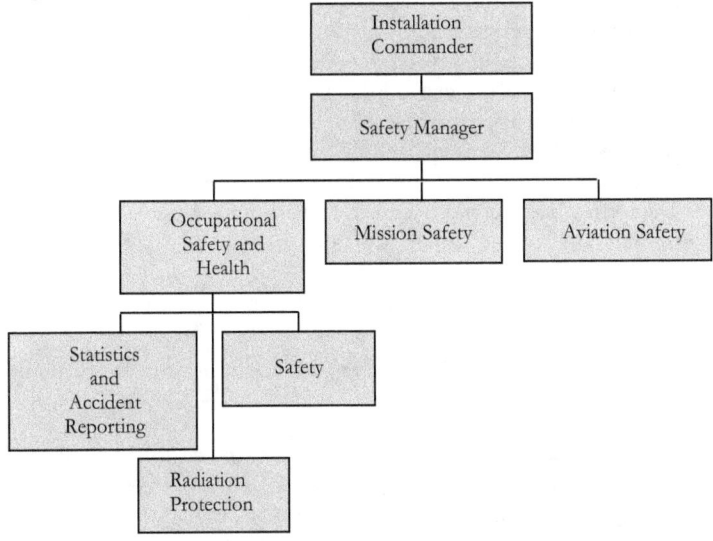

Figure 1 – Typical Safety Office Organization.

Environmental Protection

Within the duties of the Environmental Protection Program are certain ones that directly relate to public safety as opposed to just pure protection of the environment. It is important to look at the structure of this program. Army Pamphlet 10-1 (1994) is a great resource to identify the offices responsible for the environmental protection program. There is an Assistant Secretary of the Army for Installation, Logistics, and Environment and a Deputy Assistant Secretary of the Army for

Environmental, Safety, and Occupational Health. Army Regulation 200-1 states that the Assistant Secretary of the Army for Installation, Logistics, and Environment has the primary responsibility for the Army's military environmental programs. The Deputy Assistant Secretary of the Army for Environment, Safety, and Occupational Health carries out those responsibilities. The responsibilities are outlined in Table 4.

- Develops Overall Army environmental policy, guidance, and direction and serves as the primary point of contact with the Secretary of Defense, Congress, other agencies and components.
- Appoints Army representative(s) for inter-service and interagency environmental committees.
- Conducts, in coordination with Assistant Secretary of the Army (Research, Development and Acquisition), an annual review of Army environmental research and development efforts.
- Provides representation on the Overarching Integration Process Team Army System Acquisition Review Council Coordination Team.
- Provides recommendations to the Milestone Decision Authority regarding program environmental requirements and manages the Army's Defense Environmental Restoration Account.
- Serves as Department of Defense Executive Agent for selected Secretary of Defense programs and provides supervision and program oversight of the Army Environmental Policy Institute.
- Acts as co-chair with the Assistant Chief of Staff for Installation Management for the HQ Army Environmental Quality Control Council.

Table 4 - Duties of the, Assistant Secretary of the Army for Installations, Logistics, and Environment.

The Installation Commander also plays an important role in the Environmental Protection program. The duties for this commander are at Table 5. The main purpose for protecting the environment is so it will exist for the future. Another purpose is to prevent exposure to personnel and the public to toxic chemicals. We have all heard of Love Canal, Times Beach, and many other horror stories.

- Comply with Federal, state, and local environmental regulations and requirements of environmental permits.
- Appoint an environmental coordinator and provide adequate staff to support the Army Environmental Program.
- Plan, develop, budget for and execute Pollution Prevention Plans.
- Identify environmental requirements in the Environmental Program Requirements Report.
- Organize and chair the installation Technical Review Committee/Restoration Advisory Board.
- Identify and report environmental requirements that affect readiness or mission requirements.
- Identify state and locally applicable environmental requirements and execute to protect the environment.
- Integrate sensitive activities into the installation environmental program.
- Sign permit applications, permits, compliance agreements, and orders.
- Participate in the regulatory development process when proposed state or local legislation affects the installation.
- Report any criminal indictment or information or enforcement action.
- Train installation personnel to perform their jobs. Ensure maintenance of training and/or certification records.
- Require appointment and training of environmental compliance officers at appropriate organizational levels.
- Report regulatory enforcement actions and spills through command channels.
- Investigate regulatory enforcement actions, complaints, spills/releases, and correct systemic problems.
- Maintain a public affairs program supporting the Army's environmental protection and enhancement activities.
- Refer inquiries from Congress concerning environmental matters through command channels to Headquarters, Department of the Army.
- Submit required environmental reports and conduct an annual internal environmental compliance assessment.
- Ensure environmental criteria are incorporated into all new and existing construction projects.
- Coordinate and assist all installation and tenant environmental activities to ensure compliance.
- Ensure activities and tenants incorporate environmental compliance requirements into all contracts.
- Apply for and maintain all Federal, state and local environmental permits for tenants of the installation.
- Develop and implement a program to track hazardous materials and hazardous waste from cradle-to-grave.

Table 5 - Requirements of Installation Commander for Environmental Protection.

Environmental protection is a large overarching program that involves everything from the water we drink to the items we spill. This program must involve military personnel, civilian employees, and contractors to make it work. The elements of an environmental protection program are outlined in Table 6. These programs involve prevention of pollution and waste to ways of managing waste and in some cases recycling. Some elements of these programs lend themselves very nicely to prevention programs that are based on public safety. After all the air, we breathe and the water we drink are very important to public safety.

Water Resources Management Program	Oil and Hazardous Substances Spills	Hazardous Materials Management	Hazardous and Solid Waste Management	Air Program
Clean Water Act	Spill Prevention	Polychlor-inated Biphenyl (PCB) Management	Waste Minimization	Air Pollution Prevention
Recreational Water Act	Clean Up	Storage Tank Systems	Conventional Explosive Ordnance	Clean-Up or Emissions Program
		Lead Hazard Management	Medical, Dental, and Veterinary Supplies	
		Community Right to Know	Resource Conservation and Recovery Act	
			Municipal Solid and Hazardous Waste	

Table 6, Part 1-Elements of the Installation Environmental Protection Program

Environmental Noise Management Program	Asbestos Management	Radon Reduction Program	Pollution Prevention	Environmental Restoration Programs
Detection	Detection	Detection	Detection	Environmental Restoration Program
Control	Control	Control	Control	Defense Environmental Restoration Program
				Environmental Restoration of Formerly Used Defense Sites

Table 6, Part 2-Elements of the Installation Environmental Protection Program

With these two programs (safety and environment) it should be easy to see that they have a lot in common. These commonalities start at the Secretary of the Army level. By reviewing Figure 2, the similarities are obvious. The Deputy Assistant Secretary of the Army for Environmental, Safety, and Occupational Health is a Senior Executive Service level employee who has control over both programs. They begin to fragment from that level to a completely different structure at the installation level as is shown in Figures 1 and 3.

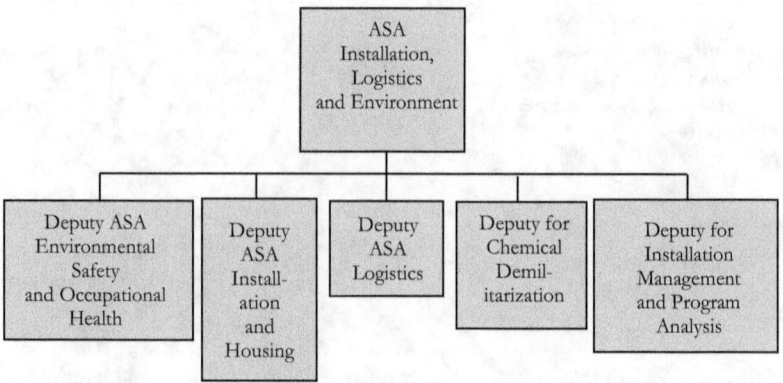

Figure 2 - Assistant Secretary Army (ASA) Structure for Safety and Environment.

Figure 3 is an organizational diagram that shows a typical Environmental Protection Program under the Director of Public works at the installation level.

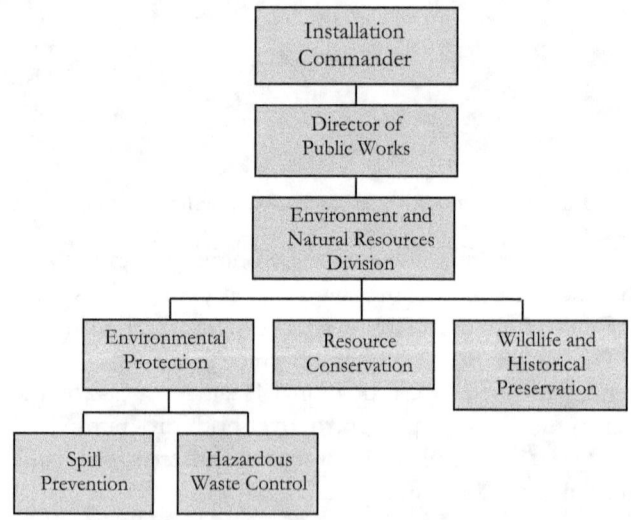

Figure 3 – Typical Environmental Protection Office Diagram

Medical

There is a Medical Command that controls all medical resources. Army Pamphlet 10-1 (1994) provides information on this command.

At each installation there is a Medical Command or hospital that provides medical support. The commander of that activity normally serves as the Director of Health Services for the Installation Commander. Within each hospital is an Occupational Health Clinic that provides community health information on a variety of medical subjects. These often include smoking cessation, high blood pressure, diabetes, stress, communicable diseases as well as sunburn protection and sports injuries. There is no typical organization for this clinic.

Military Police

Military Police organizations normally are a part of the installation structure. They usually work for the Garrison Commander. The commander of the military police unit also serves as the Provost Marshal. Crime prevention and physical security are the duties that best fit into the prevention category, see Table 7. The duties assigned to a military police unit are wide and varied, see Table 8. Army Field Manual 3-19.30 (2001) contains the Army's doctrine for physical security and crime prevention. Chapter 1, section 1-1 of Army Field Manual 3-19.30 (2001) addresses the issues of reduced resources and the impact on physical security. It states that reductions in manpower and funding are critical challenges to physical security. The field manual goes on to note that manpower for supporting physical security activities is reduced through deployments and cutbacks. This area is ripe for synergy.

- Crime Prevention Working Groups
- Crime Prevention Officers
- Crime Prevention Program Development
- Training
- Civilian Crime Prevention Organizations

Table 7 – Installation Crime Prevention Programs

Element	Examples
Systems Approach	Protective Systems, Systems Development, Integrated Protective Systems, and Security Threats.
Physical Security Lighting	Planning Considerations, Principles of Security Lighting, Types of lighting, Wiring Systems, and Maintenance.

Table 8, Part 1-Elements of the Physical Security Program

Element	Example
Design Approach	Design Strategies, Protective Measures, Vehicle Bombs, Exterior Attack, Standoff Weapons, Ballistics, Forced Entry, Covert Entry and Insider Compromise, Surveillance and Eavesdropping, Mail and Supply Bombs, and Chemical and Biological Contamination.
Protective Barriers	Fencing, Utility Openings, Other Perimeter Barriers, Security Towers, Installation Entrances, Warning Signs, Other Signs, Installation Perimeter Roads and Clear Zones, and Arms-Facility Structural Standards.
Electronic Security Systems	Design Considerations, Interior and Exterior Use Consideration, Alarm-Annunciation System, Software, Interior and Exterior Intrusion Detection Sensors, Electronic Entry Control, Application Guidelines, Performance Criteria, Data Transmission, and Closed Circuit Television for Alarm Assessment and Surveillance.
Access Control	Designated Restricted Areas, Employee Screening, Identification System, Duress Code, Access-Control Rosters, Methods of Control, Security Controls of Packages, Personal Property, Personal Vehicles, and Tactical-Environmental Controls.
Lock and Key Systems	Installation and Maintenance and Types of Locking Devices.
In-Transit Security	In-Port Cargo, Rail Cargo, Pipeline Cargo, and Convoy Movement.

Table 8, Part 2-Elements of the Physical Security Program

Element	Example
Inspections and Surveys	Hazard Identification
Security Forces	Types of Security Forces, Authority and Jurisdiction, Personnel Security, Security Clearance, Organization and Employment of Forces, Headquarters and Shelters, Execution of Security Activities, Supervision, Uniforms, Vehicles, Firearms, Communications, Miscellaneous Equipment, and Military Working Dogs.

Table 8, Part 3 – Elements of the Physical Security Program

Figure 4 is the diagram of a typical Military Police organization that provides support at the installation level.

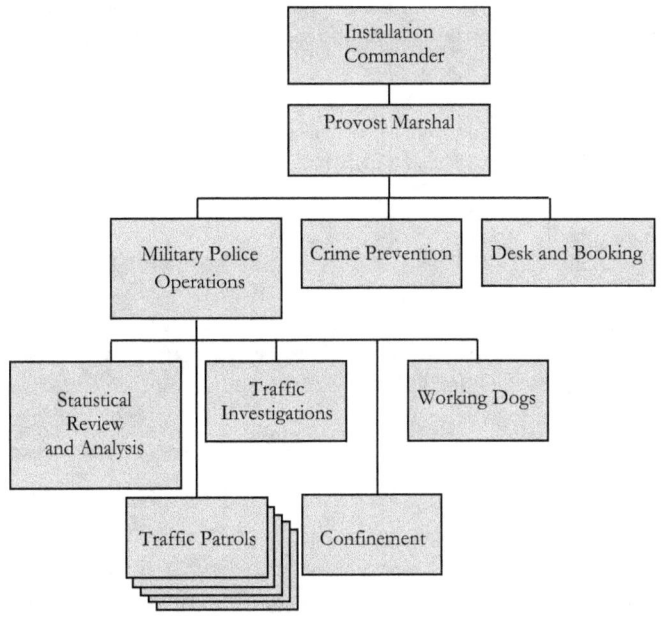

Figure 4 – Military Police Organization Diagram

This organization has the duties outlined in Table 7 and maintains the programs in Table 8. Since physical security and crime prevention are under the Military Police, they tend to be seen by the public as less significant than policing actions by Military Police. This is a shame since these programs can prevent the need for policing actions. One would assume that is where the effort would be placed; however, when budget cuts are made they normally occur in prevention programs.

Fire Department

Army Pamphlet 10-1 (1994) is a great resource to identify the offices responsible for the Fire Prevention Program. The Fire and Emergency Services comes under the Assistant Secretary of the Army for Installation, Logistics, and Environment. The Assistant Secretary of the Army for Installation, Logistics, and Environment provides policy and program direction. The Fire Department normally is part of the installation structure and works for the Director of Public Works. Table 9 shows the duties of the installation commander for fire and emergency services. The Fire Department is normally broken down into sections that protect, prevent, and respond to special circumstances.

Execute, maintain, and enforce an effective fire and emergency services program as outlined in Army Regulation 420-90, to include local and remote service activities.
Implement an information management system, such as Fire Information Management System, for use as a management tool for data maintenance and record keeping.
Conduct and approve an installation wide fire and emergency services risk analysis prior to any down-sizing actions.
Ensure that serviced tenant activities reimburse installations for fire and emergency services as defined by inter-service support agreements.

Table 9 – Installation Commanders Responsibilities for Fire and Emergency Services

The typical fire department organization involves a department with the ability to prevent fires; train personnel; fight fires; maintain sprinklers, alarms, and deluge systems; spills, hazardous material, and confined entry emergency response; and public awareness campaigns to raise the awareness of the public to what causes fires. Figure 5 shows this organization.

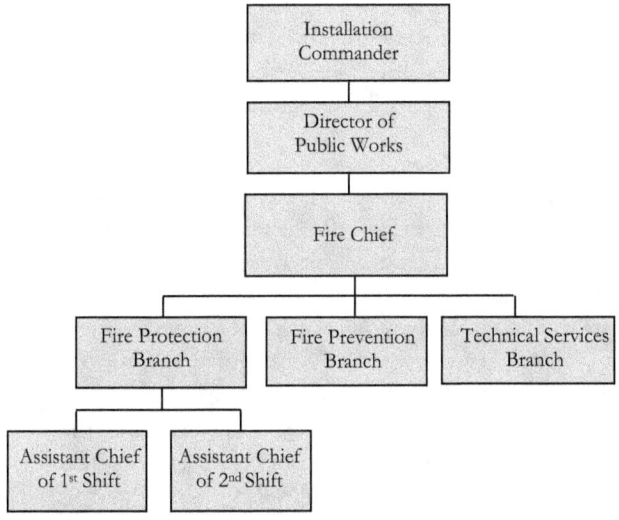

Figure 5 - Fire Department Wiring Diagram.

Chapter 6 of Army Regulation 420-90 (1997) describes the fire prevention program. The elements of this program are listed in Table 10. The installation commander is responsible for formulating a fire prevention program per appendix B, Enclosure 2 of this regulation and National Fire Protection Association (NFPA) Standard Number 1.

Development of standards for prevention at specific locations.
Implementation of those standards and education of personnel about those standards.
Conducting inspections to the standards and making hazards known.
Follow-up to ensure hazards are being corrected.
General education of population about what causes fires and how to escape from a fire.
Implementing child fire safety programs on the installation directed at school age children using, Sparky the fire dog, and Smoke Safe House.
Identifying hazards by risk assessment codes to ensure the worst hazards are corrected first.

Table 10 – Elements of an Installation Fire Prevention Program

Professional Thoughts

There is another element to this review that must be taken into consideration. That is to hear from and listen to what the professionals who work these programs think. For this review, the thoughts of a

Provost Marshal, Safety Professional, Environmental Engineer, Firefighter, Physical Security Specialist, and an Environmental Protection Specialist were solicited. Their views are summarized here and integrated into the analysis.

Part of this merger would be an alliance with law enforcement personnel to transfer the prevention functions of physical security and crime prevention to a combined organization or a merger in the missions. Lieutenant Colonel Christy Samuels (Personal Communication, October 10, 2002), Provost Marshal, had experience with an integrated approach to public safety. She indicated that it does not work due to the complexity of the broad spectrum of the work required. Her experience seemed to be a bad one. From a Physical Security perspective, the same information seems to be true. Staff Sergeant Gee (Personal Communication, November 11, 2002), Physical Security Specialist, indicated that there might be some cost savings to the merger of the many offices, but that the challenges of combining the programs would not provide any real reduction in work.

This new process would also require some changes and action from the installation safety office and personnel from within the Environmental Protection Office. Carter Boggess (Personal Communication, November 24, 2002), Safety Professional, stated that he could see a great deal of benefit in combining the functions into one cohesive directorate; however, Mr. Boggess also said he did not see how anyone but a safety professional could run such an organization. Don Barnett (Personal Communication, December 8, 1999), an Environmental Protection Specialist working in a Safety Office, believed he spent too much time doing safety work and not enough doing environmental work. He saw no benefit to this merger and believed it just might be an attempt on safety's part to gain control over the environmental protection career field.

An environmental engineer and a fire protection and prevention opinion are also relevant. Jana Brooks (Personal Communication, 8 December 2002), Environmental Engineer, and worked as a Safety Engineer and Safety Professional. She saw some benefit in the integration of these programs; however, she also believed that the personalities involved would probably take away any positives to the integration. Robert Green (Personal Communication, January 13, 2003), retired Fire Chief, believed that the integration was a very viable alternative to the cuts taken in past years to all the programs. While he was Fire Chief, the Fire Inspectors were only staffed to 40%. He would transfer the fire inspectors to another organization and believed integration would work.

Analysis of the Information

My thoughts at the beginning of this research were that this was a no-brainer. It is obvious that there was an overlap and that by better using assets the Army could save money, improve delivery of this important information and at the same time get better results; however, I was open minded about what the information and data gathered during the research would tell me. The National Security Strategy of the United States of America (2002) and the National Military Strategy of the United States (1997) both indicate that there will be a resource-constrained environment for the U.S. Army in future years. With resources scarce within the Army, extra money that does exist will continue to be used for modernization and other aspects of moving towards the objective. Short resources will continue to be a reality for these public safety organizations.

There are several responsibilities that overlap within these organizations. Table 11 is a list of these programs that can be compared for overlap by looking at the offices responsible for this program.

No.	Public Safety Program	Responsible Office or Organization
1	Fire Prevention	Safety Office and Fire Department
2	Lighting and Access Control	Safety and Provost Marshal Offices and Fire Department
3	Traffic Control	Safety and Provost Marshal Offices
4	Prevention of Drinking and Driving	Safety and Provost Marshal Offices, and Occupational Health Clinic
5	Safe Driving of two and four wheeled vehicles	Safety and Provost Marshal Offices
6	Education of Bloodborne Pathogens	Safety Office and Occupational Health Clinic

Table 11, Part 1 – Overlapping Responsibilities

No.	Public Safety Program	Responsible Office or Organization
7	Prevention of Traffic Accidents	Safety and Provost Marshal Offices
8	Inspections of barracks and work places	Safety Office, Fire Department, Environmental Office, and Provost Marshal Office
9	Workplace Violence	Safety Office, Occupational Health Clinic, and Provost Marshal Office
10	Hazardous Material and Waste	Safety Office, Fire Department, and Environmental Office
11	Fire Extinguisher	Safety Office and Fire Department
12	Fire Alarms	Safety Office and Fire Department
13	Injuries from Animals and Insect	Safety Office and Occupational Health Clinic
14	Clean Air	Safety and Environmental Offices
15	Water Treatment Operations	Safety and Environmental Offices
16	Ergonomics	Safety Office and Occupational Health Clinic
17	Education and Prevention Awareness	Safety Offices, Fire Department, Provost Marshal Office, Environmental Office, and Occupational Health Clinic

Table 11, Part 2 – Overlapping Responsibilities

There appears to be more overlap between the Safety Office and the Fire Department than any other organizations. There is also overlap in hazardous material and hazardous waste. The third area of strong overlap is with prevention of traffic accidents with the Safety and Provost Marshal Offices. The overlap with the Occupational Health Clinic does not appear to be strong. Figure 6 provides us with a wrap-up of this comparison.

The two strongest areas of overlap are item 8 and 17. Item 8 is the inspection program used by all to inspect units of the installation. The second item 17, the area of education and prevention campaigns, appears to be an area of overlap that could benefit all. These are the areas where the best return on investment could be achieved and more emphasis should be placed. If we look at the individual programs and see what organizations overlap, it would look like the data at appendix B. The education programs have the same audience and the goal of preventing something from occurring.

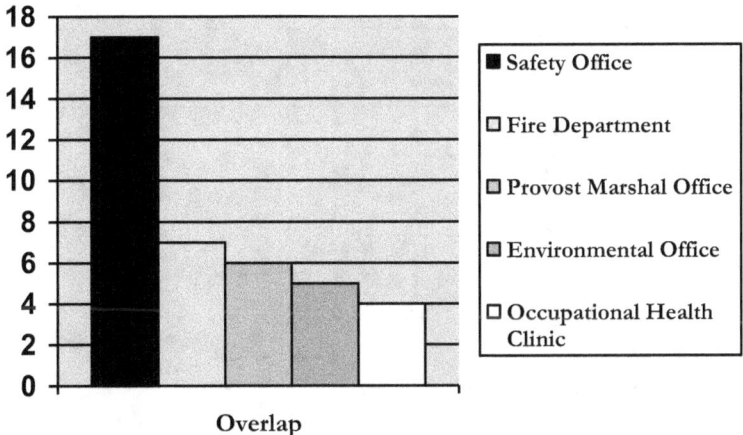

Figure 6 – Organization Responsibilities

In my book (2003), I noted that a collateral or additional duty person at the directorate and brigade level and below handles each of the responsibilities in Table 11. This makes a unified effort more important.

Despite the overlap, there is still the human factor. When I spoke to individual staff of the Army, their first impression was that I was developing a program subordinate to safety and I would be in charge. This has been done before with a very negative effect. In the 1980s, the U.S. Army Safety Center tried to obtain the missions and the resources of the Industrial Hygiene Program from the Center for Health Promotions and Preventive Medicine. This was a dismal failure that resulted in mistrust between these professionals. This aspect will have to be considered and, in my humble opinion, may be enough to prevent synergy from ever taking root.

The personnel in the career fields, offices, and organizations involved in a merger are similar and often receive the same training. A gap between them does exist, but is not impossible to bridge. We would not need to make a "super" public safety professional, but additional training is needed. This extra training may not pay off because the professionals would not be offered any additional salary for performance of the extra duties. At this point, a force field analysis is helpful. The information contained in Figure 7 tells us there are driving forces that support the change; however, it also shows very strong restraining forces that can prevent us from changing or perhaps cause failure after changes have been made.

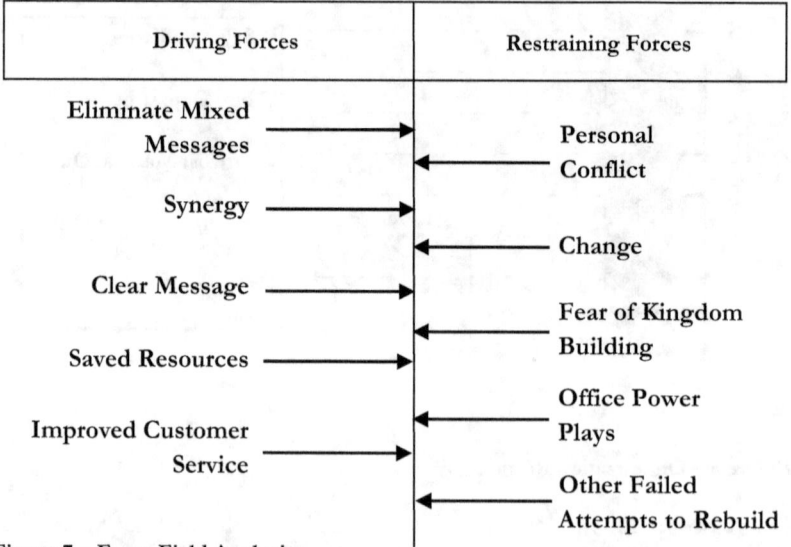

Figure 7 – Force Field Analysis

The A-MEDS Model

After identifying the driving and restraining forces, applying the Army Management Staff College-Mission Execution and Decision System will assist in problem definition and alternatives. My thesis is that limited resources and overlapping programs provide the Army with weak and inconsistent safety messages followed by an inability to provide all the services necessary and if these organizations were combined or organized to pull together, there could be one large prevention program focused in one direction for the good of all, with the overlap eliminated and resources used to fill resource shortfalls.

The known facts have been put forth up to this point. They include: work is done in a resource-constrained environment, there are some messages and responsibilities that overlap, and lastly, there is also some resistance to any implementation.

There are several alternatives to this issue. The easiest alternative was to do nothing, leaving the situation as it is and continuing with business as usual. This was the least favorable. The next possible alternative was to combine the programs of these organizations without combining the organizations. This alternative involved combining the efforts of the separate offices into a single inspection and awareness program that would coordinate the application of both programs. This was possible, but would require platform to provide oversight to ensure the program elements were coordinated. One could also combine elements of the organization under the Provost Marshal who would provide the oversight needed. This would involve moving prevention aspects of all programs under the Provost Marshal. There is also an option to form an entirely new organization combining the elements of the organizations now doing some public safety. This is a very radical alternative and would probably cause the most fear and rejection. Further still, some of these programs that have overlap could be eliminated with the duties that did not overlap given to another office. Lastly, the entire workload could be described and a work practice statement submitted for potential contracting out to a company.

Each of these alternatives will solve the problem to some degree, except doing nothing. To make this information more applicable, it is important to determine the criteria under which the issues could be resolved. These criteria are: cost saving, elimination of mixed messages, and improved service to customer base. If we measure the impact of each of these criteria on a scale of +5 to −5 with +5 being the best, 0 being no change, and -5 being the worst, we get the results in table 12.

The "do nothing" alternative provides us with no improvement or degradation from the current situation. Combining some programs as stated above would involve combining the awareness and inspection programs. This provides a weight of +6. This is a great improvement over doing nothing. Yet as mentioned earlier, who would ensure this was done? Combining elements under the existing structure of the Provost Marshal Office would give us more improvement and yet might create a monster of an organization that may not be responsive; however, this is a very strong alternative. Forming an entirely new organization is the most radical and yet may provide the best opportunity to improve all the services and improve customer service. This is the alternative with the second highest restraining forces. Eliminating some of the programs is a quick and easy idea that saves money but does not provide a better service to the customer. That leaves the last alternative, which is to contract out the work entirely. This would involve a study to write the scope of work and would also leave an opportunity for the current organizations to bid on the work. In the case of the latter that would leave us with a most efficient organization, which would do the work with fewer resources than are provided now. There would be a great deal of opposition to this alternative and it may be stopped even before it began, even though it provides a great deal of promise. Table 12 provides a comparison of specific and known criteria.

Alternative	Weighted Criteria			Total	Restraining Forces
	Cost Savings	Improved Service to Customers	Elimination of Mixed Messages		
Do nothing	0	0	0	0	0
Combine programs	0	+3	+3	+6	-1
Combine Program in Provost Marshal Office	+2	+4	+3	+9	-2
Form new organization	+3	+4	+5	+12	-4
Eliminate programs	+1	-2	+2	+1	-3
Contract out work	+4	+2	+2	+9	-5

Table 12 - Comparison of Alternatives.

There is often no best alternative to an issue. That is truly the case here. There is a better alternative and that alternative is to form an entirely new organization combining these programs and the personnel that work them; see Figure 8 for a view of this organization. This will most definitely eliminate the mixed messages, eliminate some resource constraint by eliminating overlapping work, and provide better service to the customer.

The next step would be to implement this organization. To do that would involve a staff action. The action officer would work through the Directorate of Resource Management to get accurate cost estimates on positions and overall resources. He or she would use this data to show cost savings. She or he would coordinate the action with the Director of Public Works for the Fire Department and the Environmental Protection Office, the Hospital Commander for the Occupational Health Clinic, the Provost Marshal for the Physical Security/Crime Prevention, and the Safety Manager for safety. He or she would also coordinate this action with the Staff Judge Advocate and Inspector General to ensure that it would not violate a federal statute or regulation. She or he would also coordinate the action with the Civilian Personnel Advisory Center to get a review by the union for the workers who may be the bargaining unit.

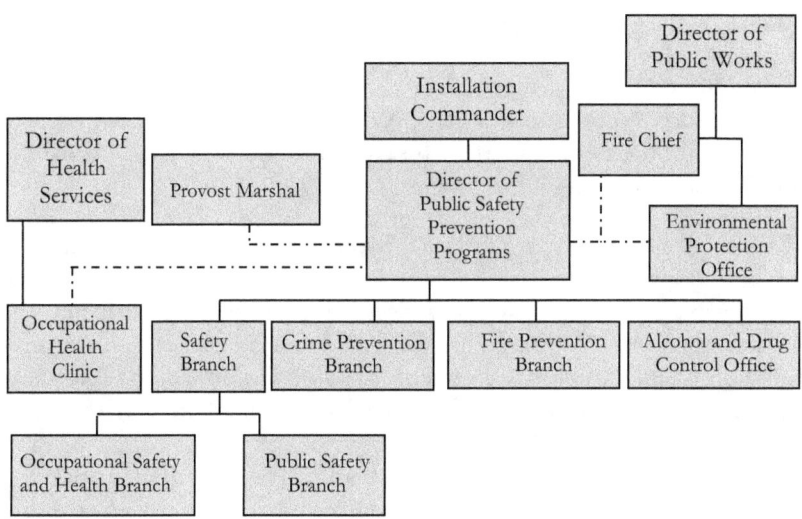

Figure 8 – Organizational Diagram of New Organization.

A new Table of Distribution and Allowances would have to be developed. There would also be job descriptions to create and others to rewrite. A new director would have to be hired from the general management series. This person should have experience with safety, physical security and crime prevention, fire prevention, and a great deal of management experience. This director could come from any of the backgrounds of the branches of the directorate.

Summary

This research project focused on increasing public safety results through synergy. This project was an analysis of the potential integration of public safety organizations that now exist within the U.S. Army. This project was a look at the many organizations of public safety that are used within the U.S. Army and their missions. This data was analyzed for the ways in which these organizations were used and the possibility that synergy may be used to save resources while at the same time providing better service to the organizations they serve.

My thesis is that limited resources and overlapping programs provide the Army with weak and inconsistent safety messages followed by an inability to provide all the services necessary, and if these organizations were combined or organized to pull together, there could be one large prevention program focused in one direction for the good of all with the overlap eliminated and resources used to fill resource shortfalls.

Furthermore, many of these offices are shorthanded, and if they could save resources through any new approach, those savings could be used to fulfill shortages of resources. Just as important was the fact that the public does not get a uniform approach to prevention.

Wiring diagrams were used to identify the current makeup of organizations. In contrast, tables were used to lay out the duties and responsibilities of these organizations. I also identified driving and restraining forces that were at work for implementing any alternative solution. The Army Management Staff College-Mission Execution Decision System or A-MEDS was used to develop alternatives and to aid in deciding upon the one alternative that would be implemented. There were several alternatives to this issue. The easiest alternative was to leave the situation the way it was and continue with business as usual. The next possible alternative was to combine the programs without combining the organizations. One could also combine elements of the organization under the Provost Marshal Office, which could provide the oversight needed. One could also form an entirely new organization that would combine elements of the organizations now doing some public safety. Elements of some programs could be eliminated. Lastly, eliminate the entire workload is yet another option, which would require describing the workload and a work practice statement submitted for contracting.

Any of these alternatives would solve the problem to some degree, except in the case of doing nothing. To make this information more applicable, it is important to determine the criteria under which the issues could be resolved. These criteria were: cost saving, elimination of mixed messages, and improved customer service. The alternative arrived at is to form an entirely new organization that would combine all these organizations and programs into one focused effort.

Bibliography

Army Field Manual 3-19.30, Physical Security. Retrieved on January 8, 2001 from
http://www.adtdl.army.mil/cgi-bin/atdl.dll/fm/3-19.30/toc.htm

Army Pamphlet, 10-1, 14, Organization of the U.S. Army, June 15, 1994, Government Printing Office. Retrieved June 2001 from
http://www.army.mil/usapa/epubs/pdf/p10_1.pdf.

Army Regulation 200-1, Environmental Protection and Enhancement. Retrieved on February 21, 1997 from
http://www.army.mil/usapa/epubs/pdf/r200_1.pdf

Army Regulation 385-10, Army Safety Program. Retrieved on March 29, 2000 from
http://www.army.mil/usapa/epubs/pdf/r385_10.pdf

Army Regulation 420-90, Fire and Emergency Services. Retrieved on September 10, 1997 from
http://www.army.mil/usapa/epubs/pdf/r420.Series.Collection_1.html.

Fanning, Fred. "Basic Safety Administration: A Handbook for the New Safety Officer, 2nd Edition" American Society of Safety Engineers, 2003, USA.

National Military Strategy of the United States of America. Joint Chiefs of Staff, 1997, Washington, DC.

The National Security Strategy of the United States of America, U.S. White House, September 2002. Washington, DC.

Appendix A - Research

The research for this essay will focus on increasing public safety results through synergy. This research will analyze the integration of public safety organizations that now exist within the U.S. Army into one organization or program.

The research will look at the many organizations of public safety used within the U.S. Army and their missions. The ways in which these organizations are used and the possibility that synergy may be used to save resources while at the same time providing better service to the organizations they serve will be analyzed.

Research will be conducted in the Maneuver Support Center Technical Library as well as a search of this topic on the World Wide Web. Information gathered in this search will be used to develop an initial lay-down of information. Searches will be conducted to determine situations in which these offices may be combined into one organization and determine how that would work. Six personnel will be interviewed to determine how this application could be used and if they believe it will save resources.

The goal is to spell out in this research that there are many public safety organizations that are focused on preventing accidents, injuries, crimes, and alcohol and drug use. These organizations are normally used without any real coordination between them. This often means their customers do not receive the kind of well-focused and direct public safety programs that provide for the safety of personnel, property, and equipment My thesis is that limited resources and overlapping programs provide the Army with weak and inconsistent safety messages followed by an inability to provide all the services necessary, and if these organizations were combined or organized to pull together, there could be one large prevention program focused in one direction for the good of all with the overlap eliminated and resources used to fill resource shortfalls.

Appendix B – Responsibilities by Number

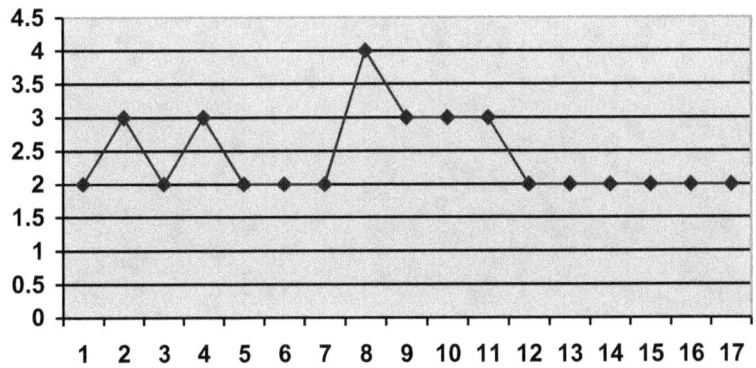

This chart depicts the frequency of overlap. The x axis is the number of occurrences while the y axis is the number of responsibility, listed below.

No.	Public Safety Program
1	Fire Prevention
2	Lighting and Access Control
3	Traffic Control
4	Prevention of Drinking and Driving
5	Safe Driving of two and four wheeled vehicles
6	Education of Bloodborne Pathogens
7	Prevention of Traffic Accidents
8	Inspections of barracks and work places
9	Workplace Violence
10	Hazardous Material and Waste
11	Fire Extinguisher
12	Fire Alarms
13	Injuries from Animals and Insect
14	Clean Air
15	Water Treatment Operations
16	Ergonomics
17	Education and Prevention Awareness

Note: An abridged version of this article was originally published in the Perspectives Newsletter, Volume 8, Number 2 in 2009 by the American Society of Safety Engineers, Council on Practice and Standards.

Essay 23-Credentials for Today's Safety Professional

Introduction

Years ago, I was approached about obtaining a credential to practice a trade. I was not quite sure what the intent of the credential was, but I knew it would take a lot of work to get. Later, I found out that it not only meant more work, it also meant that I would pay a lot of money to obtain it. My first experience was through an apprenticeship program to obtain a journeyman's license. Since then I have found a great many opportunities to obtain credentials in the workplace. Credentials can hinder or improve your ability to do your job by taking time from your family, late night studying, and some can help you earn more money. What are credentials?

What are Credentials?

Credentials come in many shapes and sizes. If you thought they were just certifications, you wouldn't be wrong but you would be limiting the definition. Webster's New Collegiate Dictionary defines credentials as "something that gives a title to credit or confidence." You may remember when career fields began requiring college degrees then degrees with specific majors. Fire departments use training and testing to determine when a person can use the title inspector, engineer, and chief. Safety, Health, and Environmental (SHE) jobs are no exception.

There is also an unofficial, but real meaning of credentials. They are often used as discriminators between people applying for a specific thing, e.g., a job, promotion, or training. Credentials are often referred to as gates. The use of degrees, specific training, and certifications fall into the same category. Those of us in the SHE careers field went through at least one gate or we would not be employed now. That was the gate of being qualified, reviewed by human resources staff, and hired into a specific position. You may have also gone through a second gate when you went for and got that first promotion. Then when you became a supervisor, you were required to complete more training.

Now we hear about obtaining advanced degrees and becoming certified just to name a couple of the more talked-about gates. Sometimes it seems we are continually meeting requirements that allow our supervisors, managers, and customers to have confidence in our ability to do our job. I have come to accept this as the way the world of work is and will always be. So, the second choice is whether I want to participate in it or not. I believe that each of us must answer this question.

What happens if I don't care to complete any of the advanced requirements or credentials? The answer is vague. There are still people being promoted that have not completed these requirements; however, I think we will see less of that in the future. The bottom line is that none of us must play the game and go through the gates any more than we decide to; however, that means that we may not receive all that is due us. We must be able to prove two things: that we are a better employee than we were and that we are better than the next employee. One way to prove this is with a credential or certification.

How to Know

Let us look at certifications. There seems to be three important questions that each of us must ask ourselves about going through the process of becoming certified.

> *What do I owe myself as a person?*
> *What do I owe my organization as an employee?*
> *What do I owe my profession as a member?*

Your answers to these questions will help you decide if you want to put yourself through the work and expense of a certification. To answer the first question, you must decide where you want to go with your career. You will then need to look at your organization and determine what role certification plays in eligibility for promotions, lateral transfers, or perhaps special training programs. Lastly, do you participate in your profession and how is that affected by certification?

If you decide that certification is not what you want and you are happy with your knowledge and experiences in the career field, that is an acceptable answer. If you're not sure how you feel about this matter, you should discuss it with your supervisor or a colleague who is certified. The process doesn't stop here; you must go on to answer the other two questions. What will be required from your organization? This should be easier to answer than the first question. If you are satisfied in your present job and your employer is happy with you, the answer may be that you don't need to be certified. If you look at your organization as a larger entity, you must not forget that things change and you may want a promotion later that may be easier to get if you are certified. I don't believe certification will be a requirement for all jobs within the next ten years. Unfortunately, that is not the issue here. The question is whether hiring officials will begin to trust the certification to provide them with higher quality employees? It is already beginning to happen. Just last week I heard about a safety person missing out on a position apparently because the hiring official used the certification as a discriminator to help him choose between two highly qualified candidates. This is not the first time I have heard of this occurring. The use of a certification for this purpose is an acceptable practice and is something you must consider when deciding whether you want to be certified.

What about your obligation to the profession? Not everyone is interested in practicing in the SHE profession. Some people see themselves as competent and making a significant contribution to their organization, while others are interested in getting promotions, better positions, and contributing outside their organization for the profession at large. For this person, I think a certification is a must. Please don't confuse the latter person as being better than the first. We all have different priorities and goals in our lives and it takes both kinds of people to get the job done.

What certification should you get? This question is about as ambiguous as all the others I have asked. It really depends on you. I sent off for packets from each of the certifying agencies and did a little research to find out what my general peer group thought of them. I also did a little research on the job market outside my organization where certifications were often required. I then decided how much money and time I had available to spend for certification. I made my decision based on these answers. I went through all the gates and was certified. Years later I changed my mind and put out a little more money and effort and obtained what I thought was a better certification, better for me. Why did I change my mind? Over the years I had grown professionally, was better trained, and had completed an advanced degree. These put me in a better position to upgrade my certification to fit my needs. I am telling you this so you will see that, in my case, it was a personal decision to become certified. Furthermore, the decision is not final and certifications can be changed or discontinued. I have colleagues who have changed their certification and others who have discontinued theirs, and in each case they were satisfied with the decision. I believe they were satisfied because it was their decision to make. One lesson I have learned from my own experiences and those of others is if you don't want to be certified, do not go through the process. It will not be worth your time and money.

As for the organization's role in certification, I think they are there to serve as our mentor and guide to help us improve not only ourselves, but also our career field. You have a right to disagree with this position; however, I think the career field must improve each year to meet the new demands placed on it. Should we require all personnel in the SHE career field to be certified? I don't believe that we should. I also don't believe that we will. It is simply not necessary as a minimum standard. There are a lot of positions that should be certified and some in fact are. I have seen positions that advertised for gas free engineers and for conducting HAZWOPER classes. Each of which asks for a certification. I also believe that the employing organization should provide us with training that gives us an advantage, so that when and if we decide to test for certification, we will pass the examination. In addition, supervisors should consider incentive or cash awards for employees who are able to complete a certification regardless of which type. This could be for the extra effort in raising the competency of the profession.

Summary

That takes us full circle. Becoming certified is still a personal choice that must involve consideration of your needs, your organization's requirements, and your commitment to your profession. Your answer is your expression of who you are and your place in the career field and each of us should respect the decision you make. One point to remember is if you choose not to be certified, you should keep up with changes in SHE through periodic training.

Note: This essay was published in the Safe Today Newsletter in the November/December edition of 1997 by the United States Army Safety Center.

Essay 24-The Art of Leadership and the Science of Management

Background

The battle rages over the difference between leadership and management. "By now most of us are getting tired of the endless attempts to distinguish between the concepts of leaders and managers. Some people have given up and use the terms interchangeably," (Murphy, 2008). Despite the abundance of disagreement, there have been many informed descriptions and definitions over the years that reflect the thought processes of the writer more than the actual meaning of either word. The original question often morphs into tirades of which word is more important, which is often followed by tirades over how to become one, the other, or both. Sadly, this battle will continue long after this essay is read, but in this essay the writer would like to add to the discussion by providing his input from over twenty years of leading and managing programs and people.

Defining Leadership

What is leadership? The American Heritage Dictionary (Leadership, 2008) defines it as:

- The position or office of a leader: He ascended to the leadership of the party.
- Capacity or ability to lead: She showed strong leadership during her first term in office.
- A group of leaders: They met with the leadership of the nation's top unions.
- Guidance; direction: The business prospered under the leadership of the new president.

Dr. Robert Murphy defines it as "A set of activities directed at influencing an individual's action within an organization with the aim of achieving organizational goals in an efficient manner" (Murphy, 2008).

Those definitions don't appear to be very helpful. Perhaps it would help to find the definition of leader. Unfortunately, the American Heritage Dictionary defines leader as "a person or thing that leads." This might be the root of the problem; no good definition of leader or leadership is easily found that is agreed upon to facilitate further discussion. The author prefers Don Clark's definition of leadership, which is "the process of influencing people while operating to meet organizational requirements and improving the organization through change (The Art and Science of Leadership, 2008).

To build on that definition, the next question is "What do leaders do?" Ms. Rosalyn Carter, former first lady, explains that "leaders take people where they want to go and great leaders take people where they don't necessarily want to go, but ought to be" (ArcaMax, 2008). The use of influencing is obvious in these statements and a very important part of leading. "When employees internalize organizational goals as a part of their own value system (private acceptance), the individual who influenced them to do so has become their leader. In the case of the leader, he or she goes a step beyond and gets the members of the group to take on the goals as a part of their own value and operating system" (Murphy, 2008).

What do Leaders Do?

Don Clark's definition demonstrates that leaders have a specific organizational context to leading, and Rosalyn Carter supports the definition by explaining that within the organizational context leaders take people places. Where do leaders take people? For a computer company, it could be to obtain market share. For a safety office, it might be reduced accident experience with minimized severity of those occurring. It is different for each organization. Julius Rhodes, an HR consultant, notes that the most important element of taking the people is that the leaders ask for commitment not compliance and transformation not transactions from the led (Presentation, 2008). Commitment and transformation require something almost spiritual from the individual, as if he or she is giving part of themselves to follow the leader. Another important point made by Julius Rhodes is that leaders often define success as "serving others" (Presentation, 2008).

Leaders also follow. Many believe that leading and following are two different things, but they are not. Part of the learning process of becoming a leader is to learn to be a good follower. It is in the position of follower that a future leader learns his or her best lessons. Just as importantly, a leader must have followers; without them there is no opportunity to lead. If a leader oversees an organization and some members of that organization choose not to follow, they can prevent or disrupt the leadership efforts of the leader.

Management

What is management? The American Heritage Dictionary (Management, 2008) defines it as:

- The act, manner, or practice of managing; handling, supervision, or control: *management of a crisis; management of factory workers.*
- The person or persons who control or direct a business or other enterprise.
- Skill in managing; executive ability.

Management may also be described as a set of activities directed at an organization's resources - human, financial, physical, and information - with the aim of achieving organizational goals in an efficient and effective manner (Griffin, 1999). According to Dr. Robert Murphy "In my opinion, Griffin includes all the elements one would expect to find in the definition of management" (Murphy, 2008).

Are Leaders Born or Made?

Are leaders born or made? The author first heard this question in high school when the football coach talked to us about being born winners and born leaders. In the coach's mind, winners and leaders were one in the same. This may not always be true, because in some cases losers often show more leadership than winners. Unfortunately, the question remains. Are leaders made or born? Leadership appears to be about the mind and heart. In this context, leaders are not born, but leaders must be born with the capacity to become leaders. That is, born with the temperament and intelligence necessary to be a leader. The leader's life then takes over and his or her values and beliefs are molded during childhood and adolescence. While in these stages of life, the leader learns the skills needed to be a leader. Watch two children playing in a park and it is easy to identify who is leading whom. In high school and at work, there are formal and informal leaders practicing the art of leadership on each other, fumbling their way through.

Summary

Let the battle rage over the difference between leadership and management. As a safety, health, and environmental professional, it is important for you to learn the steps to be successful in both leading and managing. In most cases you won't have authority and will be left to influence those you want to act, also known as leading. This essay can help by providing you with a familiarization of both leadership and management.

Bibliography

ArcaMax, Classic Quotes by Rosalyn Carter. Retrieved on October 9, 2008 from http://www.arcamax.com/quotes/s-30716-221320.

Leadership, American Heritage Dictionary. Retrieved on October 9, 2008 from http://dictionary.reference.com/browse/leadership.

Management, American Heritage Dictionary. Retrieved on October 9, 2008 from http://dictionary.reference.com/browse/management.

Murphy, Robert, "Strategic Leadership vs. Strategic Management: Untying the Gordian Knot." Retrieved on October 16, 2008 from http://www.au.af.mil/au/awc/awcgate/awc-ldrt.htm.

Presentation "Want to Be a Better Leader?...Learn How to READ" conducted by Julius Rhodes, SPHR to the ASSE Leadership Conference, October 2-4, 2008, Lombard, IL.

The Art and Science of Leadership, Glossary of Leadership Definitions. Retrieved on October 9, 2008 from http://www.nwlink.com/~donclark/leader/leaddef.html.

Essay 25-You Haven't Heard of PechaKucha?

Introduction

Have you heard about PechaKucha? Me neither, which is surprising because it is only the hottest thing in presenting. This method was invented at an architectural firm "because architects talk too much! Give a microphone and some images to an architect - or most creative people for that matter - and they'll go on forever! Give PowerPoint to anyone else and they have the same problem" (PechaKucha, 2010). An avid PechaKucha user, Sunny Marie Hackman says that it is pronounced: Paw-Chalk-Ahh-Cha (Hackman, 2010). I don't really see how she gets that out of the two words, but I'll trust her. The PechaKucha web site explains that it "draws its name from the Japanese term for the sound of "chit chat" (PechaKucha, 2010). "PechaKucha Night was started in Tokyo in 2003 as an event for young designers to meet, network, and show their work in public. It has turned into a massive celebration, with events happening in hundreds of cities around the world" (PechaKucha, 2010). "After its invention in Tokyo it went from Bern, to London, to San Francisco in 2006" (PechaKucha, 2010). Events called "PechaKucha Nights" are held around the world. They "are informal fun gatherings where creative people get together and share their ideas, works, thoughts, holiday snaps - just about anything really, in the PechaKucha 20x20 format" (PechaKucha, 2010).

The Technique

We have all experienced the long, boring, and drawn-out PowerPoint presentation and wished there was a "better way." One option for that "better way" seems to have arrived. The PechaKucha web site explains that "it rests on a presentation format that is based on a simple idea: 20 images x 20 seconds. It's a format that makes presentations concise, and keeps things moving at a rapid pace" (PechaKucha, 2010).

I was always taught that an average slide should take from 60-120 seconds. I have always tried to stick with that rule, but I can't get my mind around 20 seconds per slide. In addition to the short time per slide, the slides change automatically. I have tried that in the past using PowerPoint with little success. The idea of slides changing automatically every 20 seconds seems a little daunting; however, think of how short the human span of attention is and it does make sense. With a robust and aggressive slide show like this, the listener has little time for his or her mind to wander.

To the question "What has this got to do with training?" I answer: a lot. Trainers are always looking for ways to keep the audience's attention and leaving them wanting more. This just may be a way to do that. "It is a great format for project reviews and presentations at schools or internal presentations in offices. We are setting up PechaKucha Learning and PechaKucha Corporate programs. We also license the event format for Events and Conferences, please check out PechaKucha Event for more details" (PechaKucha, 2010). It seems to be accepted and used by the younger generation, and therefore may be a tool to gain the attention of young students. The reasons it is liked by the younger generation may be the exact reasons the older generation may be slow to accept it. Other techniques that I have tried that are fast paced and cutting edge don't seem to work well with the older generation; they expect things the way they used to be. In contrast, I have also found a great many older students who appreciate a fresh approach to training and students like that would enjoy this method.

"Say what you need to say in six minutes and 40 seconds of exquisitely matched words and images and then sit the hell down. The result, in the hands of masters of the form, combines business meeting and poetry slam to transform corporate cliché into surprisingly compelling beat-the-clock performance art" (Pink, 2010).

Application

Sharon Bowman in her book "Preventing Death by Lecture" tells us that we have our life adjusted to focus in eight-minute segments (Bowman, 2003). We do this because most of us watch television, and television shows us eight minutes of show, then commercials, then back to the show. Sharon says that we spend an average of two hours each day in front of the television and during that time "we watch 15 eight-minute chunks of information" (Bowman, 2003).

The PechaKucha method uses 6 minutes and 40 seconds to make your point. That is well within the eight-minute attention span explained by Sharon Bowman. This method could be perfect for:

- pithy pocket training,
- tailgate briefings,
- short presentations to be followed by discussions,
- introducing or summarizing training, or
- for presentations to board meetings.

It is important to have a variety of ways to communicate your point to a variety of audiences. Some refer to this as having tools in your toolbox. PechaKucha is one of those tools I recommend be in every Safety, Health, and Environmental professional's tool box.

Summary

Sunny Marie Hackman says it best: "PechaKucha is a fun and engaging way to present ideas and interact with an audience. I will use it again. My experience with the form put me on a path of continued growth as a speaker, because it taught me that experimenting with different techniques and technologies increases my skill and enlarges my territory as an effective oral communication" (Hackman, 2010). With most of us trained to such a short attention span, getting a message across in the shortest amount of time possible is essential. Your audience won't listen if you don't meet their expectations for being short and to the point. PechaKucha is just one method to get your audience to listen to your message.

Bibliography

Bowman, Sharon. Preventing Death by Lecture, 2003, Bowperson Publishing, Glenbrook, NV USA.

Hackman, Sunny Marie, What is PechaKucha? Toastmaster, Volume 76, Number 4, Mission Viejo, CA, 2010

PechaKucha Web Site. Retrieved on April 8, 2010 from http://www.pecha-kucha.org/.

Pink, Daniel, Pecha Kucha: Get to the PowerPoint in 20 Slides Then Sit the Hell Down, Wired Magazine: Issue 15.09. Retrieved on April 8, 2010 from http://www.wired.com/techbiz/media/magazine/15-09/st_pechakucha.

Essay 26-Motorcycle Safety Foundation Method of Training

Introduction

Riding a motorcycle is just like riding a bike, isn't it? I am sure as Safety, Health, and Environmental professionals, we have heard that a time or two in our career. Some adults believe this and often injure or kill themselves because of not knowing how to ride a motorcycle. There is a program that can take an adult and make them a safe motorcycle rider in two days. This program was developed by the Motorcycle Safety Foundation and is normally run by a state office of motorcycle safety or training. The author is a graduate of the Basic Rider SM, Experienced Rider SM, and the Motorcycle RiderCoach Preparation SM Courses. He also taught the Basic Rider SM and Experienced Rider Courses SM. In this essay, the author will outline this program and how it is conducted. This essay is meant as a description of the Motorcycle Safety Foundation's (MSF) program to make safety trainers aware of it.

The Bottom Line

Motorcycle riders must be trained to ride a motorcycle so they can correctly respond to emergency situations that otherwise would most likely result in an accident. Furthermore, motorcycle riders must learn about the basic clothing and equipment needed to ride a motorcycle safely and basic preventive maintenance checks to keep the motorcycle running properly. First, it is important to note that the MSF curriculum recognizes that the students are adults and come to the training with some knowledge of the subject. Secondly, the trainer serves as a coach. These two principles serve as the basis for everyone to move forward as a team with one goal, to provide the student with the knowledge and experiences to ride a motorcycle properly and safely. There are normally two RiderCoaches and a range aide for each class. The range aide sets up the range for the various exercises. This person is not a qualified RiderCoach.

Classroom Training Methods

Each session starts with a short video or discussion among the students with the RiderCoach acting as a facilitator. There are cards that guide the discussions and each student is encouraged to share his or her thoughts and experiences on the subject. Hands-on training is provided whenever possible, even in the classroom. For example, when discussing the different types of motorcycles, many RiderCoaches will provide an example of each model and allow the students to walk around and find the different controls on each model and see just what the differences are. Another example is when discussing the motorcycle protective equipment. Many RiderCoaches will have various kinds of protective equipment on hand for the students to see and feel. Many RiderCoaches will ask a group of students to pick out the proper items for a trip in cold weather. This is a great method in that the members of the group pick out items and explain why they chose those items to the class.

Dunsmore and Hausmann describe a guided discussion as one where the trainer presents initial questions or concepts but allows group learners to examine the topic. This method is useful when a trainer is trying to develop in her students the ability to assess a situation and "think on their feet" (Dunsmore and Hausmann, 2006). This is a skill each motorcycle rider needs. To properly use this method, the RiderCoach must know how to guide the discussion to achieve the learning objective.

RiderCoaches even add action to videos or after speaking. The RiderCoach will show a video or may speak on a topic for part of the class and then has students answer review questions on the material covered. The RiderCoach uses the review questions and answers provided with the curriculum. He or she may place numbers on little sticky notes under the chairs of the students before class starts. After the video or short discussion is completed, the RiderCoach asks the students to get up and stretch and look under their chairs. Under each chair is a sticky note with a number on it. The RiderCoach tells the students they only must answer one question each, the one that corresponds to the number on their sticky note. He or she also encourages students to write down the answer to each question as the other students say them. The RiderCoach calls on the student with the number one to stand up and read his or her answer. This continues until all questions have been answered. This method requires a little preplanning on the part of the RiderCoach as well as an investment in sticky notes. The RiderCoach must ensure that the notes are placed under seats that are used and make sure the numbers correspond to the review questions. The method allows students to get up and move around. It also helps students recall the information as they hear it explained in different ways by the other students.

A second method to liven up a video or after speaking might include the RiderCoach having students answer review questions on the material covered in the lecture using a list of questions from popular culture. This list can include questions from television shows, music, current events, or sports. After the video or discussion is completed, the RiderCoach asks the students to get up and stretch and then get ready to test their knowledge on pop culture. The RiderCoach then tells the students they only must answer one question each. They will be selected to answer a question by a classmate who answers a question about pop culture. The RiderCoach also encourages the students to write down the answer to each question as the other student says it. The RiderCoach reads the first pop culture question and asks the students if they know the answer. A student raises his or her hand and gives an answer. Having answered the question correctly, the student gets to pick another student to answer review question number one. This continues until all questions have been answered. This method requires a little preplanning on the part of the RiderCoach to write the pop culture questions. The RiderCoach must ensure that the answers to the questions are known to the students. The method allows students to get up and move around. It also helps students recall the information, as they hear it explained in different ways by the other students.

Also, included in the classroom portion of the training are short discussions on the dangers of riding while intoxicated, drugged or tired. These are very important subjects and the discussion can get very lively. Furthermore, classroom training does not have to be done in a classroom. RiderCoaches use the outdoors, tents, and lawn chairs to get the point across. Sessions using videos are normally done in a classroom.

Motorcycle Riding Hands-On Methods

The curriculum provides a well-developed method for allowing the students to ride the motorcycles. At first, the student learns the preventive maintenance checks and starts the motor and turns the motor off. Then the student moves to engaging the clutch and putting the motorcycle in gear and engaging and disengaging the clutch with the motor running.

The rider is now ready to go. He or she starts the motorcycle engine, engages the clutch, and with both feet on the ground, lets the clutch out just enough to move the motorcycle while walking. This progresses through a serious of actions that lead up to the rider pulling his or her feet up onto the foot pegs. What does this phase provide the rider? Obviously, it seems very basic; however, it takes the rider to the very first thing he or she needs to know about riding a motorcycle: it is motorized. This information is crucial to further learning. Each future exercise will begin the same way so the student knows how to turn on fuel, start engine, use clutch, etc.

The student continues through a series of exercises intermixed with classroom activities to ensure the learning is continuous and smooth. It is through hands-on learning that the student's muscles learn how to perform the act. This may be a new concept for many, but it is key to the success of the MSF method. The actual performance of an act is remembered at the cellular level of the human muscle, and by repeating this action, the learning is reinforced.

As the students ride the motorcycles, the RiderCoach provides input, recommendations, and encouragement. The RiderCoaches use the sandwich method to provide input. This method involves giving a compliment or word of encouragement, then pointing out the needed improvement, followed by a compliment or word of encouragement. The whole process is very positive. The negative or improvement comment is sandwiched between the two compliments or words of encouragement, allowing the student a better chance of hearing what was said rather than shutting down from criticism.

Evaluations

At various times through the classroom instruction or hands-on instruction, there are evaluations. There is also a final examination and hands-on evaluation that each student must pass; however, this comes as no surprise since the whole course has led up to these two evaluations.

The hands-on evaluation is done riding the motorcycle through events that demonstrate important fundamental riding skills. There is a little pressure from the RiderCoach to get this right, but the real pressure usually comes from the students themselves.

The riding hands-on evaluation is obviously the harder of the two and requires the student to demonstrate their skill level on the specific tasks. If the student can't pass the riding portion, then he or she cannot ride a motorcycle well enough to ride on the streets or obtain a license.

Summary

Riding a motorcycle takes knowledge and practice. The MSF training program can take an adult and make them a motorcycle rider in two days. This essay described the program to make safety trainers aware of this type of program and was based on the author's personal experiences as a student and RiderCoach. It is important for safety trainers to be aware of the many methods and approaches to training that exist.

Bibliography

Dunsmore, Scott and Paul Hausmann, Effective Training Technology Choices: Improving your ROI, *Proceedings of the 2006 ASSE Professional Development Conference*, 2006, Des Plaines, IL, USA.

References:

Motorcycle Safety Foundation, 2009. Retrieved on August 24, 2009 from http://www.msf-usa.org/.

Essay 27-The Toastmaster Method of Training

Introduction

To improve our ability as trainers, we must be exposed to a variety of training methods and styles. Through each exposure, we learn a little more and can improve our own training ability. It is for that reason that the author is a member of his local Toastmasters International Club. He has experienced firsthand the training method used by Toastmasters to make members better public speakers. "Dr. Ralph C. Smedley launched the first Toastmasters Club back in 1924" (Smedley, 1988). Dr. Smedley founded an educational organization to help its members improve communication and leadership skills. "The education program is the heart of the Toastmasters Club. It provides members with a proven curriculum that develops communication and leadership skills one step at a time" (TI Education, 2009). In this essay, the author will briefly address the curriculum and how the teaching methods used are achieving outstanding results. The author has learned a great deal from this method and believes that all trainers can benefit from learning more about the Toastmaster's International Training Method firsthand by participating in the process.

The Bottom Line

The Toastmasters International Education curriculum is focused on the adult learner. Back in 1924, no one was focusing on the adult learner by adapting teaching and methods to his or her needs. That is except Dr. Ralph Smedley. Dr. Smedley developed a club that allowed its members to practice public speaking, listening, and critical evaluation techniques in a comfortable atmosphere where everyone was equals. He did this for three very good reasons.

1. The first result of speech training is self-discovery.
2. Real communication is impossible without listening.
3. We gain skill by practice, and we improve by heeding our evaluators. (Smedley, 1988)

145

Self-Discovery

As trainers, we know that adults have a great deal of knowledge about a subject before they ever set foot in the classroom. How can we use that experience and knowledge to help the student connect with the learning material we want them to learn? Toastmasters has developed a great technique in the club. The club meets periodically and visitors are encouraged to attend. When a visitor attends, he or she is immersed in the club activities of speaking, listening, and evaluating. They see firsthand the learning process in practice. They are then asked to join and, if they choose to, they become a member. As a new member, they are afforded the opportunity of a mentor to help them with the first task of public speaking. That task is to decide to complete the Competent Communicator manual. This manual is broken down into ten speeches that give the newcomer as well as the old hand an opportunity to practice a variety of speeches. The manual provides an explanation of each speech and learning objectives. The first speech is called the "Ice Breaker." It is simply the new member speaking about himself or herself.

Listening

What on earth has listening got to do with speaking? A significant part of public speaking is the ability to speak clearly and properly. To do that, we must listen to how words are used, how voice is adjusted, and how often filler words or mispronunciations occur. This is not to grade the speaker, but rather for the listener to hear how others speak and identify those same weaknesses or strengths in him or herself. New words and even new meanings can be learned through listening. In addition, the listener can learn through listening how other members' speeches are delivered and speak in the same fashion when it is his or her turn. Listening is also key to proper evaluation, and in the club method, everyone has an evaluator to provide feedback.

Evaluations

There is no instruction in a Toastmasters Meeting. Instead, members evaluate one another's presentations (How Does It Work, 2009). "In Toastmasters, feedback is called evaluation, and it is the heart of the Toastmasters educational program" (Effective Evaluation, 2009). An evaluator is assigned to listen and critique each speech. Each speech is given an oral critique in front of the whole club and then a more detailed written critique in the member's manual. There is also a grammarian, Ah Counter, and timer who will provide the speaker feedback about the speech. The speaker gains a lot of feedback on his or her presentation and a chance to improve and speak again soon. All critiques are done with the best interest of the speaker in mind. No one is there to demean or belittle a speaker, but rather to build them up. No one tells a member what to speak on and no one passes or fails a speaker, yet improvement occurs. That is because all members are there for the same reason: to improve their public speaking skills. This works because of the club and cannot be learned in isolation (Communication Track, 2009).

Each member takes a turn at evaluation so that he or she can learn the skills of effective listening, developing feedback, and then if feedback. All of which are valuable life lessons in addition to good speaking lessons. The evaluator provides an honest reaction in a constructive manner to the speaker (Effective Evaluation, 2006). To help the evaluator, Toastmasters International provides evaluation guides, but the evaluation is clearly the evaluator's opinion and nothing more.

All speakers need feedback about their attempts to speak effectively, and without it, speakers are speaking with blinders on. The evaluator gives the speaker the information he or she needs to improve their speaking ability. It is still up to the speaker to choose what advice to take. This is where multiple evaluations come in handy. The speaker hears from a different evaluator after each speech, and from these sessions he or she starts to see a bigger picture of what the audience sees and hears. This is valuable information.

Summary

To improve your ability as a trainer, you must be exposed to a variety of training methods and styles. Through each exposure you learn a little more and can improve your own training ability. It is for that reason that the author is a member of his local Toastmaster International Club. He has experienced firsthand the training method used by Toastmasters to make members better speakers. Since 1924, Toastmasters has been providing an educational program to help its members improve communication and leadership skills. "The education program is the heart of the Toastmasters Club. It provides members with a proven curriculum that develops communication and leadership skills one step at a time" (Education, 2009). In this essay, the author briefly addressed the curriculum and how the teaching method achieved outstanding results. The author has learned a great deal from this method and believes that all trainers can benefit from learning more about the Toastmasters International Training Method firsthand by participating in the process.

References:

Effective Evaluation, Toastmasters International, 2006, Toastmasters International, Inc., Mission Viejo, CA, USA.

Smedley, Stanley, PhD, Personally Speaking, 1966, 1988 Toastmasters International, Inc., Santa Ana, CA, USA.

Education Program Toastmasters International (TI). Retrieved on August 18, 2009 from http://www.toastmasters.org/MainMenuCategories/WhatisToastmasters/CommunicationandLeadershipTraining.aspx.

Communication Track Toastmasters International (TI). Retrieved on August 18, 2009 from http://www.toastmasters.org/MainMenuCategories/WhatisToastmasters/CommunicationandLeadershipTraining/CommunicationTrack.aspx.

How Does It Work? Toastmasters International (TI) Retrieved on August 18, 2009 from http://www.toastmasters.org/MainMenuCategories/WhatisToastmasters/HowDoesItWork.aspx.

Essay 28-Accessible Training and Presentations

Introduction

It may not sound very innovative to make training accessible for all students who attend, but the truth is that this is not normally done, and any effort put forth to facilitate learning is indeed innovative. This method can be used with little or no effect on the students in your class who do not experience accessibility issues. You must make all your training accessible, and in this essay I will provide important information about how you can make training inclusive so all can learn.

Hearing Impaired

In a Professional Development Conference Proceedings Paper, Sharon Campbell (2000) reminded us that making sure the student gets what we give also applies to those students who have issues with hearing, seeing, or learning. Sharon states that the biggest problem is that many students are either in denial about their communication difficulties, or are unaware of them. As the trainer, you can use innovation to overcome some learning difficulties and ensure the students get the learning he or she came for. Two good rules of thumb for any safety trainer are to "1) assume that some of your audience members cannot hear well and 2) never offer anything verbally that isn't also available simultaneously visually" (Campbell, 2000). Accessible training and presentations are those in which the trainer or speaker has considered the possible realm of communication issues that his or her students may be experiencing and have made accommodations to assist the student in overcoming the issues and learn the material. Every student coming into the classroom deserves a chance to learn and using accessible training can give them that chance. Sharon Campbell (2000) said it best: "Always remember the purpose of safety training, and realize that if someone is only partly trained because of a failure to fully understand the material, the consequences can be severe...particularly if you were asked and refused to accommodate a request for assistance." The benefits far outweigh the potential cost and inconvenience of providing these services to students. There are many examples of these methods around us, but many don't recognize them. Trainers should use videos that are open captioned or closed captioned with the captioning turned on to ensure students can hear or read the material being presented. Respond to all requests to do something to help a student hear. Use an assistive listening microphone or stationary microphone and don't walk away from the stationary microphone. If the audience is over 100 people, the trainer can use real time captioning or sign language interpreters. There are also assisted listening devices that can be used by students to ensure they receive the material. Technical terms should always be defined to ensure that every student understands what the trainer means by them. Last but certainly not least, test students to identify their retention and level of understanding of the material (Campbell, 2000).

Literacy Method

As the trainer plans for any training, he or she must bear in mind that 10 million adults in this country are illiterate in the English language (Copeland, 2003). Furthermore, 30% of Hispanic workers are illiterate in their own language (Copeland, 2003). In her 2003 presentation in Denver, Laura Copeland recommended that handout information be at the 6th grade level. Trainers must ensure that all training materials can be understood by their students, who have a variety of reading levels. This means that graphics can be used instead of words so that less reading is required. More of the learning can be demonstrated rather than written. Videos can also be used to explain learning objectives rather than having students read. All of this is done to make sure the student can learn. This will not be without some extra cost and work for the trainer. The key is for the trainer to keep the learning interesting so that the students who can read well are not turned off by the material or presentation method. The payoff is that the entire class can meet the learning objectives without leaving anyone behind. The person left behind has a higher chance of causing an accident.

For example, this method can be used by a trainer who has a learning objective for his students to know the basic steps in a Bloodborne Pathogens program. The trainer begins his instruction by showing a film that shows the basic elements put forth in the Code of Federal Regulations and the contents of a first aid kit visually rather than in writing. The film is 10 minutes in length and should be easily understood by each of the ten students in the class. The trainer then asks the students to recall the individual steps and define them for the other students of the class. The trainer passes around two kits for the students to see as the instructor holds up an item and calls on a student to name the item and tell what it is used for. The class, with video and demonstration, takes 50 minutes and allows for 10 minutes of questions. After questions, the instructor ends the class.

Bilingual Training

In 2001, Latino workers represented 11% of the labor force in this country, but they also represented 17.1% of the workplace injuries or illnesses that resulted in lost workdays. One of the main ways to improve worker safety is by communicating with employees in the language they understand. Trained employees were more successful in demonstrating basic knowledge of workplace safety and health concepts; however, they do not demonstrate significantly more knowledge than those that have not received training. "Oral Presentation in Spanish and Bilingual Training Methods allows the effective understanding of concepts, participants' interaction in the classroom with an effective scrutiny, and knowledge, of key terminology in English" (Ruano and Sanchez, 2005). Lizzette Vargas-Malpica notes that there is growing interest among governmental agencies in developing safety and health training and technical material in Spanish to help workers overcome language barriers (Vargas-Malpica, 2005). Conducting training bilingually can be a challenge for the trainer; however, with proper training and support, it can be done properly. There will be additional costs associated with this training for translation and development of training materials in more than one language. The good news is that all students can participate. Hispanic workers often avoid asking for clarification or explanations to prevent embarrassment over their limited English language ability (Vargas-Malpica, 2005).

The trainer of a respiratory hazards class speaks Spanish naturally and English is his second language; however, he speaks it fluently. The instructor provides the instruction in Spanish and has all his handouts in both Spanish and English so that his students can learn the English words they will see in the workplace compared to the Spanish words they are familiar with. The trainer begins the class of Spanish speaking students with a Spanish speaking video that introduces the topic. The trainer then goes on to a hands-on demonstration with clear plastic bags of dust, particles, and fibers so each student can specifically see the hazard. The trainer follows this with an explanation of respiratory protective devices and notes the name of each in Spanish and English and refers the students to their handouts for the actual words in written form. The instructor answers any final questions and gives a quiz in Spanish with terms in English and Spanish.

Older Workers

All around us we read and hear how Americans are older than ever before, and that as the Baby Boomers retire, there will be a shortage of workers that will require many older workers to continue working. Aging influences the ability of older workers to learn; however, with minor accommodations, older students can learn and retain that learning as effectively as younger students. The methods used for older workers include *brain-based learning* and *situated learning*. Brain-based learning is low stress in a collaborative environment. It is filled with a multitude of tasks that occur as life occurs at the job site. This method focuses on twenty-minute blocks of time to maintain focus. The whole learning experience is enriched by multimedia. Situated learning takes place in the social and physical environment so that students can learn from each other in a real setting (Jackson, 2005). For older students, it is important to reduce the amount of tasks that require the use of memory. This is done by providing the student with handouts and take-home material for them to refer to later. Ensure the classroom or training area is well lit with limited background noises to impair the hearing. Handouts should be on bright white paper with black letters for sharp contrast (Arditi, 2008). The font style is also important, and Dr. Arditi recommends a Roman font with a 12 pitch, with standard letter and line spacing (Arditi, 2008). This will not be without some extra cost and work for the trainer. The key is for the trainer to keep the learning interesting so that students of all ages can learn without some feeling left out. The payoff is that the entire class can meet the learning objectives without leaving the older students behind. The older student left behind has a higher chance of causing an accident, and with the healing response of the older body taking longer than a younger body, this employee may be out of work longer.

Say a trainer is conducting a class on the OSHA accident reporting requirements for supervisors of a local construction company. The group has several older employees who are wearing reading glasses in the class. The trainer notices that one employee in the back has a hearing aid. The trainer begins with six overhead slides to list a few basic changes. These slides have as few words as possible with 28 pitch fonts that can easily be read from the back of the classroom. The trainer also uses a white slide with black letters to make the letters easier to read. As he speaks, the trainer uses a pin on the microphone so all the students can hear her. She breaks the class up into four groups and gives each group a situation and questions for the group to answer. The handout is on white paper with 12 pitch fonts which is easy to read. Being in the groups allow the students to discuss the situation and respond to the questions, allowing all members of the group to participate. The final assignment of the class is for each student to complete two example forms. These forms are on bright white paper with Times New Roman-12 pitch font so they are easily read. Each student completes the task to standard. The trainer answers any remaining questions and ends the training.

Summary

Even after reading this essay, you may not be excited about making training accessible for all students who attend; however, I hope you will take every opportunity to facilitate learning for all your students. These methods are very important to students in your class who have accessibility issues. Take the information in this essay to heart and make all your training accessible.

References:

Arditi, Aries. Making Text Legible: Designing for People with Partial Sight, 2008. Retrieved from URL http://www.lighthouse.org/accessibility/legible/ on February 11, 2008.

Campbell, Sharon Lynn, *Accessible Training and Presentations*, Proceedings Paper, American Society of Safety Engineers Professional Development Conference, 2005, New Orleans, LA.

Copeland, Laura, Training that Rocks, Proceedings Paper, American Society of Safety Engineers Professional Development Conference, 2003, Denver CO.

Jackson, Alma, *Health and Safety in an Aging Workforce*, Proceeding Paper, American Society of Safety Engineers Professional Development Conference, 2005, New Orleans, LA.

Ruan, Norman and David Sanchez, *The Importance of Bilingual (English/Spanish) Workplace Safety and Health Training: Methodologies*, Proceeding Paper, American Society of Safety Engineers Professional Development Conference, 2005, New Orleans, LA.

Vargas-Malpica, Lizzette, *Training in Occupational Safety and Health in Immigrant Communities Tailored to Cultural Backgrounds*, Proceeding

Fred E. Fanning

Paper, American Society of Safety Engineer Professional
Development Conference, 2005, New Orleans, LA.

Essay 29-Classroom vs. Computer Based Training

Introduction

The author was asked to compare classroom and computer based training to be used to modify behavior for a traffic court. This was motor vehicle-related training for drivers that lost their driver's license and were attending court appointed training as one means of getting their driver's license reinstated. The courts were investigating the potential for cost savings of drivers going online and completing the training instead of attending four hours of face-to-face classroom training. The author taught the old Defensive Driving Course face-to-face in the 1980s for over three years, and in the 2000s served on the acquisition team reviewing bids for an online computer based version of a defensive driving course. He also taught the Motorcycle Safety Foundation's rider safety courses face-to-face for over two years.

The Bottom Line

Classroom training, although normally more expensive because of the classroom, provides a better learning environment that supports the student and allows for security to ensure that the offender takes the training instead of an imposter. The cost of a classroom can be reduced considerably by using facilities that already exist and are used for other purposes. See Table 1 for a comparison of both methods.

Classroom Training

Classroom training provides an opportunity for students to ask the trainer questions on the topics as they come up. It also provides an opportunity for students to hear answers from questions asked by other students. Students can work in groups that allow for ideas and thoughts to transfer from fellow students, which can facilitate the learning since they come from peers. Students can also speak with the trainer during breaks to address issues they feel uncomfortable speaking about in a class. Classroom training makes good use of sensory learning methods of seeing, hearing and doing. Accountability is present because the student is in class and his or her identity can be checked. Finally, classroom training adapts to all students with or without computer training or access to a computer.

Computer Based Training

Computer based training provides an opportunity for students to ask the trainer questions through e-mail. This method also provides an opportunity for students to read answers from questions asked by other students through chat rooms. Students can work in groups that allow for ideas and thoughts to transfer from fellow students, which can facilitate the learning since they come from peers using chat room and e-mail strings. Students can e-mail the trainer at any time to address issues. This method makes good use of sensory learning methods of seeing, hearing, and doing the material presented by the instructor. This method also adapts to students with access to a computer as well as computer training or experience.

Topic	Computer	Classroom
Resources	Costs of computer and internet connections.	Cost of classroom and audiovisual equipment.
Student focus	Student centered through electronic connection.	Student centered through face-to-face experience.
Sensory learning methods	Use sight, listening, and hands-on doing.	Uses sight, listening, and hands-on doing.
Access	Not all offenders may have access to a computer.	All offenders would have access to classroom.
Prior learning required	Students need to know basics of computers to include e-mail and web program operations.	None.
Security	Uses passwords but no guarantee that another person won't take course for student.	Student is present and ID can be checked to verify. Exams are controlled.

Table 1 – Classroom versus Computer Based Training

Findings

Classroom training, although normally more expensive because of the classroom, provides a better learning environment that supports the student and allows for the security of ensuring that the offender is the one taking the training. The cost of a classroom is normally reduced by using facilities that already exist.

If the students could complete the training at home on a computer, there is a question of validating who the student is. Also, many offenders may not have access to a computer. This would put them at a disadvantage to those having a computer. To exacerbate this issue, some libraries that could provide computer access are reducing hours. There is also a question of computer literacy among the disadvantaged that would come into play with computer based training. If a center was provided that verified the student's actual identity, the costs would be like classroom based training.

The idea of training that uses three sensory modes is also very important and should not be underestimated. This supports the learning theory that adults prefer to work through information and get physically involved. It is important to know that adults, just after training or learning takes place, retain 20% of what they read, 30% of what they hear, 40% of what they see, 50% of what they say, 60% of what they do, and a whopping 90% of what they see, hear, say, and do (Copeland, 2003). Compare that to the fact that one year after training, the average adult only retains 10-15% of what he or she learned. Given the dramatic loss of information, it seems obvious that a trainer must use the method that gives the best retention, which is to provide the student with learning that allows the student to see, hear, say, and do.

Summary

In this essay, the author outlined information he gathered from comparing classroom and computer based training to modify behavior for court required training. Through his own experiences, information from adult learning, and an overview of each type of training mode, the author determined that classroom training was more appropriate to this situation. The disadvantage of computer access was also considered. The author determined that classroom training, although normally more expensive because of the classroom, provides a better learning environment that supports the student and allows for the security of ensuring that the offender is the one taking the training. The cost of a classroom can be reduced by using facilities that already exist.

This is not the end of the story and much more work needs to be done to evaluate and determine the best mix of classroom versus computer-based training, especially with trainers talking about providing training via iPod or MP3 players.

Bibliography

Copeland, Laura, "Training that Rocks," *Proceedings of the 2003 ASSE Professional Development Conference*, 2003, Des Plaines, IL, USA.

Note: This article was originally published in The Communicator Newsletter, Volume 2, Number 2 in the Winter of 2009 by the American Society of Safety Engineers, Council on Practice and Standards.

Essay 30-Project Management for SHE Professionals

Introduction

Over the years I have seen Safety, Health, and Environmental (SHE) professionals manage projects without adequate training or experience, and in many of those cases, the project failed. This was unfortunate for the both the SHE professional and the stakeholders of the project. Just as in other professions, it is beneficial to have procedures that are standardized for project management. This standardization should be followed by ensuring that all personnel involved in the project are properly trained and fully understand their responsibilities. This will lead to projects that are properly managed and successful. The good news is that the standardization for project management has been done by the Project Management Institute (PMI). This organization publishes the Project Management Body of Knowledge (PMBoK®) and extensions for the public sector, construction, etc. The PMBoK® is also an ANSI standard. In this essay, I will outline project management standards and the appropriate levels of training to familiarize the SHE professional with the basic information needed to manage a project properly. This will be based on my experiences and research of appropriate texts.

Basics

Before we jump right into project management, first let me say that there are projects, programs, and portfolios. Each is interrelated with the other and together they are very important to managing work. A project is defined in the second edition of the Combined Standards Glossary published by PMI as *a temporary undertaking to create a unique product or service with a defined start and end and specific objectives that, when attained, signify completion*. The definition your organization uses for projects may differ from this, but it is important to know what the professionally accepted definition is.

A program is defined in the third edition of the PMBoK® as *a group of related projects managed in a coordinated way to obtain benefits and control not available from managing them individually or a series of repetitive or cyclical undertakings*. The second part is what most of us would refer to as operational work. A portfolio is defined in the third edition of the PMBoK® as *a collection of projects or programs and other work that are grouped together to facilitate effective management of that work to meet strategic business goals.*

Many SHE professionals speak about managing programs in the case of the Personal Protective Equipment Program, Driver Education Program, Hazard Communication Program, etc. In these cases, they are referring to conducting operational work to accomplish a single goal and not managing multiple projects. Many SHE professionals have managed projects in the case of implementing an automated accident reporting program or Material Safety Data Sheet electronic program; however, most do it using a homegrown or personally created process. There is a better way to manage projects that gets results and is based on the size of the project to make sure that the investment in project management does not outweigh the results of the project.

Project Classification

In my experience, it is best to classify projects so that the level of effort equates to the cost of the project. I have always found it best to use three levels that begin with the basic, Level 1, and end with the most complex, Level 3.

- Level 1 Projects usually have a budget under $100,000, the project team has less than two full-time members, the completion time is under six months. Examples exist of very similar projects, and the project contains low risk and/or exposure (OHRM, 2006).
- Level 2 Projects have a higher budget between $100,000 and $300,000, a project team with two or three full-time members, a completion time between six and twelve months, a department-wide scope and impact, and the project contains moderate risk and/or exposure (OHRM, 2006).
- Level 3 Projects have a budget over $300,000, a project team with more than three full-time members; completion time exceeding one year, wide scope and impact, and the project contains high risk and/or exposure (OHRM, 2006).

Most projects that will be managed by a SHE professional are Level 1. There may be occasions for a Level 2 or 3, but these size projects don't occur very often in SHE. I have managed many projects and only once did one exceed $100,000. With the right information, most SHE professionals can manage Level 1 projects successfully.

Project Manager Requirements

It is very important that each project is led by an appointed project manager referred to as the PM. It is also important to ensure that the PM is properly trained, but not excessively trained, to save money. PMs should be trained to oversee a specific level of projects. Prior to identifying training, the PM responsibilities must be identified. The PM:

- is the individual responsible for managing the overall project and its deliverables,
- acts as the customers' single point of contact for the project,
- and controls planning and execution of the project's activities and resources to ensure that established cost, time, and quality goals are met.

I mentioned earlier that projects are best classified at three levels. PMs should also be trained to those same three levels from the novice, Level 1, to the most capable, Level 3. This ensures that the PM is working on a project they are qualified to oversee.

- Level 1 PMs must complete training outlined in the next section.
- Level 2 PMs must complete training outlined in the next section, complete a university basic level certificate program or be certified as a Certified Associate Project Manager (CAPM®) through PMI, and have acquired 12+ months experience as a PM for Level 1 projects.
- Level 3 PMs must complete the training outlined in the next section, complete a university advanced level certificate or be certified as a Project Management Professional (PMP®) through PMI, and have acquired 48+ months experience as a PM for Level 2 projects (OAS, 2006).

Project Manager Training

In the previous section, I provided the basic requirements for each level of project manager. Specific training that should be completed by each PM is outlined below based on the level of the project. Instructor-led training or computer based training can equally cover the topics necessary to ensure the PM has the knowledge to do the job properly. Several of these topic areas can be included in a single course. All the courses are part of the PM curriculum within a learning path (PMBoK®, 2004).

Level 1 project PMs should be trained in the following area:

- Planning a Project
- Project Scheduling and Budgeting
- Controlling and Closing a Project
- Project Management Framework
- Project Management Process
- Project Scheduling and Budgeting
- Project Integration
- Project Scope
- Project Time Management
- Project Cost Management
- Project Quality Management
- Human Resource Management
- Project Communication Management
- Project Risk Management
- Project Procurement Management

Level 2 PMs should have completed training in all the areas for a Level 1 PM plus training in the following areas:

- Building Productive Stakeholder Relationships
- Project Estimating Techniques
- Managing Accelerated Projects
- Project Management Maturity
- Leading the Project Team
- Communicating Within a Project Team
- Overcoming Obstacles

Level 3 PMs should have completed training in all the areas for a Level 2 PM plus training in the following areas:

- Portfolio Management
- Organization, Strategy, and Business Needs
- Navigating Corporate Structures
- Bringing Home, the Value
- Selling Project Management to the Organization

Project Requirements

Now that the project level has been identified and we know what level of PM is needed to oversee the project, it is essential to determine the requirements for each level of the project. The following requirements represent standards for each project level; however, alterations are acceptable and often necessary for unique project situations.

Level 1 projects should be managed by a Level 1, 2, or 3 PM. The minimum documentation used during the project includes the use of a status update form, an approved statement of need, a list of key stakeholders, a Work Breakdown Structure (WBS), procurement requests, and lessons learned. The bi-weekly written status report is the only report that is necessary. There are two presentations that should be done. These are the project initiation and closeout presentations (OAS, 2006).

Level 2 projects should be managed by a Level 2 or 3 PM. The minimum documentation used during the project includes the use of the status update form, an approved business case, a project charter, a requirements analysis, a communication plan, a WBS, a risk assessment, change control plan, deliverable sign-off and inspection test plan, procurement requests and contracts, and lessons learned. There are two reports that are needed that include the bi-weekly written status report and the compliance with project communication report. There are two presentations that should be done. These are the project initiation and closeout presentations (OAS, 2006).

Level 3 projects should be managed by a Level 3 PM. The minimum documentation used during the project includes the use of the status update form, an approved business case, a project charter, a requirements analysis, a communication plan, a WBS, a risk assessment, earned value management plan, change control plan, deliverable sign-off and inspection test plan, procurement requests and contracts, and lessons learned. There are two reports that are needed that include the bi-weekly written status report and the compliance with project communication report. There are two presentations that should be done. These are the project initiation and closeout presentations (OAS, 2006).

The individual project update is a scheduled time for project managers to provide a "pulse check" on the status and activities within the project to the project sponsor. Each project presentation is intended for a high-level and restricted to no more than five to seven minutes, including questions and answers. Level 3 project status should include Earned Value Management techniques with relation to schedule, cost, and scope.

Change Management Plan

For Level 2 and 3 projects, it is important to have a Change Management Plan to ensure that all changes to a project are reviewed and approved in advance, changes are coordinated across the entire project, and stakeholders are notified of approved changes to the project. Changes to projects at Level 2 and 3 can cost considerable amounts of money and time if not managed properly.

The goals of a Change Management Plan are (OHRM, 2006) to:

- Identify, define, evaluate, approve, and track changes through to completion.
- Bring the appropriate parties (depending on the nature of the requested change) into the discussion.
- Negotiate changes and communicate them to all affected parties.
- Give due consideration to all requests for change.
- Modify Project Plans to reflect the impact of the changes requested.

The Change Management process may be simple or complex and should include the following steps (PMBoK®, 2003):

- Submit change request in writing
- Review change request.

- Approve or reject change request for further analysis
- If approved, perform analysis and develop a recommendation
- If rejected, return to submitter
- Accept or reject the recommendation
- If accepted, update project documents and re-plan
- If rejected, return to submitter
- Notify all stakeholders of the change.

Change requests are reviewed daily by the Project Manager or designee and assigned one of four possible outcomes. The change can be:

- rejected,
- deferred to a future date
- accepted, or
- set aside until additional information is received.

Work Breakdown Structure (WBS)

The WBS is a list of the steps of a project that identifies the start and finish dates of each task, the amount of time each task should take, resources used by each task, and any milestones for the project (PMBoK®, 2003). Project Managers should be encouraged to use more advanced features available to identify critical paths, relationships between tasks, and tasks that are behind schedule.

Project managers document the WBS and Schedule by accurately maintaining a chart. This can be done by hand for a simple project, Microsoft Excel spreadsheet for a more complicated project, or Microsoft Project for a very complicated project.

Summary

SHE professionals are normally not trained to properly manage projects. Because of this, many have failed when trying to implement a project. This is unfortunate because PMI has developed standards and a body of knowledge to help anyone called upon to manage a project. Most SHE professionals don't need to be an expert in project management, but do need to know where to find out how to manage a project. In this essay, I outlined the project management standards and the appropriate levels of training to familiarize the SHE professional with the basic information needed to manage a project. Armed with this information, the SHE professional can now get the training needed and successfully lead a project.

Bibliography

A Guide to the Project Management Body of Knowledge, Third Edition (PMBoK® 3rd Ed), (ANSI Standard 99-001-2004), Project Management Institute, 2004.

Combined Standards Glossary, 2nd Ed, Project Management Institute, 2005, USA.

Office of Administrative Services (OAS), US Department of Commerce, Project Management Guidelines, August 2006.

Office of Human Resources Management (OHRM), US Department of Commerce, Project Management Guidelines, January 2006.

Essay 31-Safety Professionals Demonstrate the Warrior Spirit

Introduction

Different people think of public service in different ways. Some people even serve an entire career in public service and don't fully comprehend what their service meant. It is the public servant who has been given the privilege and honor of making the government of the United States work. Politicians come and go. Congress makes the laws. The courts look at things from the perspective of constitutionality. The Executive Branch executes the laws. However, the public servant is called upon to make all these laws, rules, and decisions work for the public. The Department of the Army's Safety Professionals are but one group of public servants. They are public servants who stand behind the tip of the sword, prepared to support the war fighters in the defense of the Constitution and interests of our nation. These public servants, like the war fighters they serve, are a breed unto themselves. The author believes the Safety Professionals of today's Army demonstrate a Warrior Ethos in their work and dedication. They have served in times of peace and war and through good times and bad. They suffer through budget battles and furloughs, workforce reductions and having their worked contracted out, and still they show up for work to help the war fighters defend our nation. Warrior Ethos is a culture that is normally found in military service with strong character and beliefs, which are highly regarded.

Does the Warrior Ethos espoused by the warriors of the United States Army apply to the Army Safety Professional? In this essay, the author will clarify how the Army Safety Professional demonstrates the Warrior Ethos. Furthermore, he will lay out how this ethos can and should be encouraged in the Army Safety Professionals to prepare them for the 21st Century and the Transformational Army.

Public Service

Safety Professionals within the U.S. Army are hardworking, dedicated public servants who go above and beyond the call of duty to serve the nation. You can hear from personnel within the uniformed services about officers and senior noncommissioned officers who are referred to as having the Warrior Ethos. This is most obvious in the combat arms such as the Infantry. However, many of the same people who say these things will often belittle or explain away the dedication of a civilian employee. Many of us can remember when a presidential candidate, Governor William J. Clinton, and his vice-presidential candidate, Congressman Albert Gore, went on television and explained how inept government employees were. They even pulled out stacks of regulations and expensive ashtrays as examples of that ineptness. This may have helped their campaign, but it also reinforced stereotypes about public servants. These stereotypes need to be reexamined.

Most of the public does not understand what the typical public servant does. More specifically, most of the public doesn't understand that the typical public servant does not make rules or laws but rather often is tied up or hampered by them in their effort to serve the public. These misunderstandings also apply to Department of the Army civilian employees and Army Safety Professionals. It is unreasonable for one to opine that all Army civilian employees as well as Safety Professionals are exceptional public servants. They are a cross-section of the population of America and mirror the kinds of people that are found in most cities and towns. It is after they enter public service that the change can and should take place to instill in them a dedication and focus not found in the private sector, but not all make that change as well as one would like.

The Warrior Ethos

On the web site for the US Army's Infantry Officer Basic Course is a paper entitled "The Warrior Ethos: All Soldiers Are Warriors." In this essay, the warrior ethos is defined as the embodiment to win our nation's wars despite every adversity. This paper goes on to say that this ethos is what provides a soldier with the will to win, to refuse to accept failure, and to do what is right with pride. Furthermore, this paper goes on to say that the warrior ethos fuels the soldier's ability to fight, allows him to create victory out of the battle chaos, and maintain the deadly spirit that wins battles and campaigns. Lastly, the paper lists the tight fabric of loyalty as the noblest aspect of the Warrior Ethos (The Warrior Ethos, 2002). If all soldiers are warriors, can this ethos be used in a broader sense to include the civilian employees of the U.S. Army?

As you consider this question, it may be beneficial to look at another aspect of the Warrior Ethos. Major General Honor'e and Major Cerjan

(Honor'e, 2002) provide additional explanation in their article "Warrior Ethos the SOUL of an Infantryman". In their paper, they refer to the Infantryman as the "Tip of the Spear." In this position, they opine that the Warrior Ethos is what sets the Infantryman as the model for the U.S. Army. This is a great point for you to begin to answer the question of whether the Warrior Ethos can be applied to an Army Safety Professional. In this paper, the Major General Honor'e and Major Cerjan (2002) opine that there are twelve characteristics that make up the ethos. One clear point must be made now. It is not the goal of the author to compare Army civilian employees to Infantryman or soldiers. These men and women are clearly "Warriors" and are the mainstay of the security of our great nation. Instead, it is the author's goal to look at the attributes that comprise the Warrior Ethos and determine if and how Army Safety Professionals demonstrate the ethos. Having made that point, twelve characteristics must be looked at to see if in fact an Army Safety Professional can demonstrate these characteristics.

Major General Honor'e and Major Cerjan (2002) list the twelve characteristics as:

> *self-discipline; belief that one's word is one's bond; mental toughness to endure; embodiment to guard one's post; iron will, determination and confidence to overcome; relentless desire to be the best; uncompromising commitment to be technically and tactically competent; inherent selflessness to give your last and to replace me with we; unqualified willingness to sacrifice oneself for the mission; the ability to overcome the horror of battle; to never give up; and lastly to always put the mission, unit, and the country first and oneself second.*

This is an impressive list, and one which would be hard to live up to. Hence the question, can an Army Safety Professional demonstrate the Warrior Ethos? It may help to look at the military and civilian employee cultures.

Military versus Civilian Culture

Before that question can be answered, you must look at the culture of the military compared to the culture of the civilian employee. First, can these cultures integrate or must they remain separate? Roy Eichhorn

169

believes that people can create an integrated culture among military and civilian personnel. If Mr. Eichhorn is <u>correct</u> and these two cultures can be integrated, then it is possible for the Army Safety Professional to demonstrate the Warrior Ethos. Mr. Eichhorn explains that there is no Army civilian culture, but that "The civilian culture is a function of some basic consistencies and the local sub-cultures which can reflect region, mission, or size of installation." Mr. Eichhorn further explains that the military culture as we see it is much more uniform by design. This uniformity is often lacking in the civilian culture. This creates the gap that often exists between military personnel and their civilian colleagues; however, misunderstanding makes the gap larger (Eichhorn, 2002).

The misunderstanding that Mr. Eichhorn describes is fueled in part by the uniform rules of socialization established to stabilize the uniformed member's life. In this later context, uniformed refers to service members who wear uniforms i.e., soldiers, sailors, airmen, and marines. These rules often leave the civilian employee out of the culture of the organization, causing frustration and negative feelings toward the uniformed members. This basic difference in culture revolves around the uniformed member working in a fast-paced up or out system that requires frequent moves and short-term assignments. The civilian employee, on the other hand, is normally a steadfast employee serving long-term in assignments and jobs. Many civilian employees spend an entire career moving up only two or three pay grades (Eichhorn, 2002).

The basic reason for some of the misunderstanding is beginning to melt. Some civilian employees are now moving from job to job and many serve in more than one career field in their public service. Commanders can move civilian employees into harm's way if needed.

On the Army Civilian Personnel Online website, you can find a unique section on the oath of office for civilian personnel. In the section entitled "Our Common Bond" you can read the following paragraph:

> *Welcome to the Army Team! A Team made special by the oath you and others take to support and defend the constitution of the United States. Those who took the oath and served before you consistently demonstrated the Army's leadership values of loyalty, duty, respect, selfless-service, honor, integrity, and personal courage (Oath, 2003).*

That short paragraph seems to paraphrase the twelve principles of the Warrior Ethos as described by Major General Honor'e and Major Cerjan (2002). That same website goes on to say:

> *When taking the oath, you accept the same demands now that American soldiers and Army civilians have embodied since the Revolutionary War. The oath deals with values and ethics. The acceptance of and adherence to the leadership values of Loyalty, Duty, Respect, Selfless Service, Honor, Integrity, and Personal Courage will lead to successful and rewarding careers like those of the citizen soldiers who served in the early years. These attributes are collectively referred to as the Army ethic. By instilling these values within each soldier and Army civilian, we can strengthen the professional Army ethic.*

When you consider this information, it does not seem that the gap between military personnel and civilian employees is as large as it once appeared. The main difference appears to be that the military culture is uniform across the service while the civilian culture, if it exists at all, is not uniform across the service and is dependent on the local culture of the installation or organization. The preceding comments alluded to the oath of office. What exactly is this oath? It may surprise some to know that military officers and civilian employees take the same oath. It is as follows:

> *I do solemnly swear (or affirm) that I will support and defend the constitution of the United States against all enemies, foreign and domestic; that I will bear true faith and allegiance to the same; that I take this obligation freely, without any mental reservation or purpose of evasion; and that I will well and faithfully discharge the duties of the office upon which I am about to enter (Oath, 2003).*

Real Examples

Within this perceived dichotomy of cultures, or what are better described as sub-cultures of one organization, can Safety Professionals break out of this typical mold and demonstrate a Warrior Ethos? There is the case of Dennis P.F. Woolsey. Mr. Woolsey is now retired but spent most of his career assigned as a civilian employee to tactical military units. He spent from 2-6 months in the field most years and was deployed to Iraq for Operations Desert Shield and Storm and well as Bosnia-Herzegovina with Operation Joint Endeavor. Mr. Woolsey believed in the twelve characteristics put forth by Major General Honor'e and Major Cerjan. Mr. Woolsey practiced the twelve characteristics in his selfless service; mental toughness to endure; embodiment to serve his unit of assignment in war and peace; his relentless support of his units that embodied his desire to be the best; and unqualified willingness to sacrifice oneself for the mission. Mr. Woolsey spent several tours in Germany serving in forward deployed tactical units. He also sacrificed a great deal of time with his family because of his commitment to his work. Mr. Woolsey suffered excessive stress and illness because of his demonstration of the Warrior Ethos.

There is also the case of Carter T. Boggess, Jr. Mr. Boggess spent most of his career assigned as a civilian employee to tactical military units. He served in several civilian career fields. Mr. Boggess also spent from 2-6 months in the field most years and was deployed to Hungary, Bosnia-Herzegovina, and Kuwait. Mr. Boggess believed in the twelve characteristics put forth by Major General Honor'e and Major Cerjan. Mr. Boggess practiced these twelve characteristics in his selfless service; loyalty to his organization; embodiment to serve his unit of assignment in war and peace; and his unqualified willingness to sacrifice oneself for the mission. Carter Boggess spent a great deal of time on temporary duty. In the two tours, he spent in Iraq he gave more than his fair share and made untold sacrifices for his work. He also spent a great deal of time in Germany serving in forward deployed tactical units.

Lastly, there is the case of Jonathan Foster. Mr. Foster is the son of a career Army civilian employee. Mr. Foster had a more stable career serving in tactical military units and base operations or installation organizations. While he was assigned to tactical military units, Mr. Foster was often in the field from 2-3 months a year and was deployed to Hungary and Bosnia-Herzegovina with Operation Joint Endeavor. He also served as the Safety Officer for the bridge across the Sava River. Mr. Foster is a humble public servant and would never tell you he strove to meet the Warrior Ethos; however, his actions over a career demonstrate he believes in the twelve characteristics put forth by Major General Honor'e and Major Cerjan. Mr. Foster practices these characteristics as a quite professional. He demonstrates selfless service;

while at the same time believing that one's word is one's bond. If you would have seen Mr. Foster knee deep in mud at the Sava River Crossing, you would not need to ask about his loyalty to his organization or his embodiment to serve his unit of assignment in war and peace. He gave more than his fair share and made untold sacrifices for his work.

Warrior Ethos and Army Civilian Employee

If you look at the description and application of the Warrior Ethos and compares that to an Army Safety Professional, the juxtaposition is obvious. Or is it obvious? The Warrior Ethos is clearly a military goal and this humble attempt to adapt it to Army Safety Professionals is in no way meant to reduce the esprit it builds in military organizations. However, if the U.S. Army is truly one team fighting one fight then it is only reasonable for Safety Professionals to strive, in as much as possible, for that same goal. But what does the research say?

The Warrior Ethos is adequately described in papers on two web sites. On the Infantry Basic Officer Course web site, the paper entitled "The Warrior Ethos: All Soldiers Are Warriors" defines this ethos as the embodiment to win our nation's wars despite every adversity. Major General Honor'e and Major Cerjan provided additional explanation in their article "Warrior Ethos" The SOUL of an Infantryman. These authors opine that there are twelve characteristics that make up the ethos. Roy Eichhorn believes that people can create an integrated culture among military and civilian personnel. Mr. Eichhorn explains that there is no Army civilian culture; it is more of a subculture that adapts to local regions. Mr. Eichhorn further explains that the military culture as we see it is much more uniform by design. This uniformity is often lacking in the civilian culture. This creates a gap that often exists between the military personnel and their civilian colleagues. It is the author's opinion that this information leads people to believe that it is possible for an Army Safety Professional to strive for and demonstrate the Warrior Ethos. It is possible since the Army Safety Professional can adapt to the ethos of the warrior and make that his or her culture. This means that the Safety Professional would strive to meet the twelve characteristics. In addition, the examples provided may seem extreme to some and ridiculous to others. Yet these public servants sacrificed to meet the missions of their organization. This allows them to live up to their oath of office and the ideal set by the uniform services. These sacrifices do not come without a cost. If more civilian employees were to strive for the Warrior Ethos it would mean a great deal of sacrifice, and yet isn't that what the oath and Army ethic call for?

Future Challenges

If Safety Professionals are to succeed in the 21st Century they, as well as all public servants, must adapt to a new work ethic. The Warrior Ethos is a model that can be used by Safety Professionals of the U.S. Army. This will mean that they will have a better chance of being embraced as part of the team. Furthermore, the Safety Professional must adapt to the culture of the U.S. Army as much as possible. The Civilian Personnel Online web site held some additional information of great interest. The page on leader development describes how the Civilian Training, Education and Development program supports the development of leaders. This program works to develop in leaders eleven skills which are:

- Anticipate, manage and exploit change.
- Versatile enough to operate successfully in war and operations other than war.
- Exemplify the highest professional and ethical standards.
- Uphold the dignity of everyone.
- Display technical and tactical proficiency, while exploiting the full potential of advanced technology and accounting for the human dimension.
- Possess teaching, coaching and counseling skills.
- Build cohesive teams.
- Communicate effectively while stimulating confidence, enthusiasm and trust.
- Accurately assess situations, solve problems, and act decisively under pressure.
- Show initiative, plan thoughtfully, and take reasoned, measured risks to exploit opportunities.
- Clearly provide purpose, direction, motivation and vision to their subordinates while executing operations following their superior's intent.

This training program for civilian leaders is intended to be like the training for the development of Army officers. That would lead people to believe that there may already be movement in the direction of developing the Warrior Ethos within the Army civilian employee. The performance support forms and evaluation reports for senior civilian employees are very like the same documents for Army Officers and in fact have a block for comments on values and beliefs held by the employee. This is or will start to move the civilian employee to demonstrate the competencies of the Warrior Ethos.

Summary

Can someone be an Army Safety Professional and still demonstrate the Warrior Ethos? Based on the research conducted by the author and the opinions put forth in this essay the answer is a firm yes. This does not mean that all Safety Professionals will adapt the ethos; however, it does mean that the Army is moving in that direction and at least the civilian leaders will be trained and educated in the basic elements of leadership. This will allow them to set a goal to demonstrate the ethos and make it their own. The lean and acutely focused Army Safety Professional will fit in better with his or her military counterparts if they understand and at least try to demonstrate the values and beliefs of Warrior Ethos. This will bring them closer to being one team and one fight.

Bibliography

Eichhorn, Roy; Creating an Integrated Military/Civilian Culture, Army Management Staff College. Retrieved on February 15 from https://www.amsc.belvoir.army.mil/ecampus/sblmp-nr/readings/NR_Term1_Readings/organizations.

Honor'e, Russel L. Major General and Major Robert P. Cerjan (2002, January). Warrior Ethos the SOUL of an Infantryman. Center for Army Lessons Learned Newsletter.

Oath of Office, Civilian Personnel Online. Retrieved on March 1, 2003 from http://cpol.army.mil/permiss/74b.html.

The Warrior Ethos: All Soldiers Are Warriors, (2002, 18 Sep), Basic Officer Leadership Course Website. Retrieved on March 3, 2003 from http://www-benning.army.mil/BOLC/ethos.htm.

References

Deployment Requirements, Civilian Personnel Online. Retrieved on February 12, 2003 from http://cpol.army.mil/library/911info/civ-mobil.html.

FM 22-100, Army Leadership, Headquarters, Department of the Army, August 1999.

Welch, Samuel. (2002, February). Warrior Development and the Human Side of Knowledge Management. Paper presented at the Army Knowledge Symposium 2002, Overland Park, KS.

Note: This essay received first place in the 2003 Federal Manager Association sponsored writing contest at the Army Management Staff College.

Essay 32-US Army Safety Professionals Go to Make Peace

Introduction

It has been years since the end of Operation Joint Endeavor. Yet the successes of that operation are still as valid today as they were when they helped American soldiers cross the Sava River, develop base camps in empty fields, and implement the Dayton Peace Accord in a region that experienced four long years of war. Considering recent missions now is a good time to revisit the safety program and some of the personnel that made this such a safe and successful operation.

Implementing the Dayton Peace Accord

Operation Joint Endeavor began in the fall of 1995 under the worst Balkan winter and flooding of the Sava River in seventy years. As the soldiers began to deploy, few were aware that an initial effort to reduce risks had started months before they packed their bags. Early in 1995 the V Corps Safety Office had begun the move to a deployable safety office that could support all battle staff planning and risk management as well as execute the safety program on the ground. This initiative was to play an essential part in the success of the operation.

An annex to the Operation Joint Endeavor campaign plan was written that focused on standard compliance and risk management. To support that annex, risk management was conducted of all railheads for the deployment from Germany. Units conducted risk management of their operations using both the deliberate and hasty risk assessment methods. US Army Europe (USAREUR) Safety and V Corps Safety combined efforts to develop waivers for locations to authorize uploading of ammunition in combat vehicles. These waivers were based on a risk assessment method developed by Jim Schooler of USAREUR Safety. This new method allowed for an easy to explain visual depiction of the hazards and their relationship to the uploaded vehicles.

177

Challenges and Controls

Commanders at all levels faced severe challenges caused by the extreme winter weather, flooding, and the short timeline for the deployment. In addition, the lack of any infrastructure within Bosnia-Herzegovina caused housing problems and a severe shortage of good road networks, and an overabundance of mud made travel difficult. With this extreme situation before the deploying units, the need to provide a safe working and living environment within Bosnia-Herzegovina was evident and a great deal of work was done to balance the needs of the mission with the need for hazard control. This was an extremely difficult task.

Commanders used a variety of methods to identify and control hazards. The operation worked from a performance to standard base. Standards were first identified, trained to, and then validated prior to a unit deploying to the operation. This was done through a series of exercises called Mountain Eagle and Mountain Shield. The intent was to verify through a third party, most cases V Corps staff, that the unit was in fact ready to go, and where shortcomings were identified, the validator mentored the unit through the process of improving to standard. The V Corps Aviation Safety and Standardization Detachment validated all aviation units through the entire operation. In addition, they supported incident and accident investigations. Second, Unit leadership was tasked to act as standard enforcers. This reduced the use of shortcuts and missed steps to get the job done. In addition, civilian safety personnel were a third element. They were located strategically throughout the entire theater of operation to assist the deploying units as well as the unit they were assigned to; it was a real team effort. After the initial deployment, Wartime Army Safety Personnel that were later called Army Safety Assistance Detachment (ASAD) personnel, were brought in to support the operation at locations in Taszar, Hungary; Slavonski-Brod, Croatia; and Tuzla, Bosnia-Herzegovina. The ASAD personnel were a very important element of the entire process and it would have been hard to do it without them. This is but one story in a series of great efforts. Commanders also used the Force Protection Working Group to identify and control hazards. When you compare the numbers of fatalities, disabilities, and property damage with the number of soldiers on the ground, extremely bad roads, and no fixed facilities, it is evident that commanders were successful in controlling hazards.

The structure of the operation was at times confusing and yet there was a distinct methodology that was created to support the soldier on the ground. The chain of command for Army Forces was through the Implementation Force (IFOR), the subordinate ground force was the Multi-National Division North (MND-N), and beneath that was Task Force Eagle (TFE). Task Force Eagle was the unit most U.S. Army forces were assigned to. In addition, there was an Army Force (ARFOR) headquarters in Hungary called U.S. Army Europe (USAREUR) (Forward). USAREUR (Forward) provided the title 10 support to the U.S. Forces. The Intermediate Staging Base (ISB) in Hungary was the first operation most soldiers saw as they deployed. The ISB was designed to serve as an intermediate stop for soldiers deploying to Bosnia-Herzegovina. The operation consisted of Taszar Air Base, Kaposvar Barracks, and Kaposujlak Air Base. The bulk of deploying soldiers went through the ISB operations at Taszar Air Base Life Support Area. Kaposvar served as a logistic operation and housed a good many of the (some-what) permanent party. Kaposujlak was the Air Base used for rotary wing operations and maintenance. The soldiers arrived by bus, plane, and train to Taszar. They were housed in large tents, matched up with their equipment, provided ammunition and moved to Slavonski-Brod, Croatia.

There was a second staging base set up in Slavonski-Brod with many of the same operations and hazards as in Hungary. In addition to the staging base, this area was also used for crossing points of the Sava River with the hazards inherent in that operation. From Slavonski-Brod, soldiers went into their base camps in Bosnia-Herzegovina. In some cases, soldiers went to muddy fields that would become base camps sometime in the future, after they helped build them.

Civilian safety professionals played a role from beginning to end of Operation Joint Endeavor. Initially, they were living in tents and sub-standard buildings (just like the soldiers) with no heat or plumbing, in the middle of winter. Ed Hoffman and Carter Boggess from V Corps Safety were the first two into Taszar, Hungary to create the USAREUR (Forward) Safety Office at the ISB. To support the ISB and the deployment Joe Sapp of the 3rd Corps Support Command (COSCOM) deployed to Hungary along with Mike Moody of the 29th Area Support Group. Another important member of the 3rd COSCOM team was John Cecil. John spent the better part of the year as the safety manager for 181st Transportation Battalion in Bosnia-Herzegovina and Croatia. Mike Wood and Tanya Griffea deployed with their unit the 1st Armored Division to Bosnia-Herzegovina. Mike and Tanya served as the safety office for Multi-National Division North, Task Force Eagle and the 1st

Armored Division simultaneously. Tina Whittington from V Corps Safety served a tour as the safety manager of the USAREUR (Forward) assisted by Roy Valiant from V Corps, Safety.

Rail operations were a primary safety concern. In the beginning the concern was for rail loading. One note that should be made here is that trains in Europe often run with electric engines that require overhead power lines that carry thousands of volts. When a soldier is loading, or off-loading, he or she is at risk of electric shock or electrocution. Risk assessments were completed and control measures implemented to reduce the risks for railheads in Germany. This allowed the movement of material and equipment on rails near their locations. Rail training was conducted and by Movement Control Teams set up at the different sites to provide on-site support for all rail operations. The local Base Support Battalion assisted with and in some case, did a great deal of the rail loading, another procedure that reduced the risk. The Red Cross was at most locations to provide a warm beverage, snack, and a care package to each soldier and civilian before they departed. The focus was shifted from the rail loading to rail transporting hazards. Security teams were used on the trains and two serious accidents occurred because guards used improper procedures. Focus was turned to a complete analysis of the transportation of the security teams, and new policies and guidance were developed that allowed for the completion of the initial deployment and subsequent redeployment with no further serious accidents for security guards. The emphasis then shifted to offloading the equipment from rail cars.

Convoys were the second major concern for the deployment. The narrow roads, long Main Supply Routes, and limited rest areas increased the risk significantly. In some areas, the roads were so narrow that two Army convoys could not pass each other from opposite directions at the same time. Thanks to some creative management, convoys were routed, dispatched, tracked, and accounted for so that no two convoys met in a spot they would not pass or pile up on each other at rest stops. This was one of the great success stories of Operation Joint Endeavor brought about by the soldiers of the USAREUR (Forward) Redeployment Operations Cell in Taszar, Hungary. The narrow roads, long Main Supply Routes and limited rest areas increased the risk significantly and these risks were controlled through the application of proper management techniques. Due to the serious winter weather, a plan was also developed that arranged the authority for movement of convoys from different areas in the region. Task Force Eagle supervised the area south of Main Supply Route Paul and Task Force 21 supervised the area north of Main Supply Route Paul. In both cases, specific personnel were authorized to release convoys in inclement weather. General officers made the decisions that allowed convoys to move in extreme inclement weather, e.g. black ice, heavy snow, etc. Risk management was an important part of the go/no-go decision making process for convoy movements.

The Sava River Crossing was an engineering feat that will go unmatched for some time. Jonathan Foster, safety manager, volunteered to leave his job with the 130 Engineer Brigade to serve as the safety manager for the Engineer Brigade, 1st Armored Division for the river crossing. When the 130th Engineer Brigade assumed the river crossing operation, Jonathan continued with the 130th Engineer Brigade to see the river crossing through to success and sustainment. The river crossing was exacerbated by severe flooding and international pressure to cross the river to meet the time constraints of the Dayton Peace Accord. A valuable lesson was learned about outside pressures caused by civilian authorities and news media. This outside pressure may cause a unit to succumb to higher risks to try to relieve the pressure or spare them embarrassment. The 130th Engineer Brigade had placed great importance and constant emphasis on training to standard back in Germany and that was shown through under the difficult conditions of the Sava River.

Cold weather was the third major concern. The forecast was for high levels of snow, cold temperatures, and prolonged exposure to severe weather. A plan was needed. A strong cold weather injury prevention program started before the operation, along with command emphasis, reduced injuries to an unbelievable four out of 20,000 troops on the ground through a harsh Balkan winter with little or no improved housing or work areas. V Corps Safety developed a "Winning in the Cold" booklet from books used in Korea and the 8th Infantry Division. The book was a single source document for cold weather operations, injury prevention, and equipment usage. V Corps Safety also brought in trainers from the Mountain Warfare School in Vermont to conduct training for leaders of deploying units. A major control measure was to issue additional cold weather clothing and equipment to each soldier to protect them from the harsh Balkan winter. This additional equipment included one pair of extreme cold weather boots, two pairs of cold weather boots, extreme cold weather sleeping bags, and many other items. Some equipment was issued to allow the soldier time to clean the items so that they could wear one, have one drying from cleaning, and a third being cleaned.

Mines were the fourth major concern and special training was provided in what was called STX training. STX refers to situational training exercise. This training was continued after the initial surge and renamed Individual Replacement Training (IRT). Each soldier and civilian employee received specific instructions on what the most common mines looked like, how and where they were buried, and what to do if they found a mine. The most valuable parts of the training were: what to do if you find a mine and what to do if you believe you are in a minefield. V Corps Safety also sent out a safety tape that addressed the hazards of mines, a mine safety booklet, and mine awareness graphic training aids.

STX training was a three-day training class. The students lived in a field environment which included tents and kerosene space heaters. Classes were provided on the political situation, rules of engagement, motor vehicle safety, mine safety, manning a checkpoint, conducting vehicle and personal searches, and conducting a patrol. This was an outstanding tool to educate soldiers on the force protection issues involved in the operation. The field training involved searching vehicles and personnel, manning a checkpoint, and conducting a patrol (complete with booby traps and mines). In addition, a live mine demonstration was conducted just to make a believer out of you.

During the sustainment phase, there were several other rotations supported by civilian safety professionals. Dennis Woolsey and Ed Hoffman both of V Corps Safety took a turn rotating as the safety manager of Multi-National Division North, Task Force Eagle and 1st Armored Division after Mike Wood completed his tour of duty. Steve Murane from the 7th Army Training Command also deployed to develop ranges. Jim Schooler, USAREUR Safety deployed to assist in the resolution of ammunition issues in Hungary, Croatia, and Bosnia-Herzegovina. Furthermore, Marvin Ballard from 3rd COSCOM provided additional support to 7th Combat Support Group in Taszar, Hungary. Another important member of the 3rd COSCOM team John Cecil was still there. Anne Ferguson, USAREUR Safety, worked a tour as the safety manager of USAREUR (Forward) assisted by Mark Peterson. Mark also spent a great deal of time in Bosnia-Herzegovina assisting Task Force Eagle with range design and construction as well as ammunition and explosive storage, and ammunition uploading and downloading.

Much of the work done during sustainment was focused on stabilizing the force by improving base camps and supply routes. This became a full-time job for many. The focus through all this was to meet the intent of the Dayton Peace Accord and that is where most soldiers spent their time, on checkpoints and patrols. Safety personnel worked with unit leaders, force protection working groups, and allied safety personnel to move towards a safer environment where more and more risks were controlled. The challenge was to reduce the amount of risk from hazards that were accepted while at the same time not reducing the units and soldiers' ability to conduct their missions. Mr. Woolsey provided the structure to an allied presence by holding the first Allied Safety Representative Council during this period. Safety representatives from the United States, Russia, Poland, Turkey, and the NORDIC brigades all took part.

As mission change signaled the end of Operation Joint Endeavor, preparations were made to bring in a covering force to support the withdrawal of the 1st Armored Division. The covering force was identified as the 1st Infantry Division. As we began the initial phases of redeployment, Fred Fanning took over from Anne Ferguson as the safety manager of USAREUR (Forward) assisted by Don Barnett, an Environmental Protection Specialist from V Corps Safety.

To assist with the redeployment of 1st Armored Division and the deployment of 1st Infantry Division, additional assistance was needed. The task was to provide support for the deployment of approximately 8,500 soldiers and all their equipment from Germany and the United States through Hungary and Croatia to Bosnia-Herzegovina. Right after that the task was to support the redeployment of approximately 20,000 soldiers and all their equipment from Bosnia-Herzegovina through Croatia and Hungary back to Germany or the United States. This was a monumental task. The question was "Could all this be done before another Balkan winter set in? The answer was a resounding yes! It was obvious that more safety personnel were needed. Ms. Helma German, USAREUR Safety, deployed to support the 7th Combat Support Group who was running the ISB and later to USAREUR (Forward) Safety. Gary Ziola and Rovelma Hudson from USAREUR Safety supported the 1st Armored Division in Bosnia-Herzegovina and Croatia. Rovelma even stayed to support operations at the ISB in Taszar, Hungary. Tina

Whittington served as the safety manager for 1st Armored Division and Task Force Eagle during the redeployment.

Risk management was also an integral part of the redeployment. Units from the 1st Armored Division conducted risk management of the redeployment. When units processed through the ISB in Taszar, Hungary, they updated their risk management daily to factor in the effect of change. All non-divisional units also conducted risk management. The hasty and updated risk management covered the next day's operations. The 7th Combat Support Group was the service provider for all deploying and redeploying units going through Taszar, Hungary. Under the leadership of their commanding officer, Colonel Byron Lester, they conducted risk management daily and briefed unit personnel on their findings. This risk management was compared to the results of the redeploying unit's risk management and inconsistencies were corrected. The 7th Combat Support Group is another chapter in a series of great successes.

Under the leadership of Colonel Art Floyd USAREUR (Forward) Deputy Chief of Staff for Personnel and Lieutenant General John Abrams, USAREUR (Forward) Commander, the USAREUR (Forward) Safety Office conducted daily threat assessments and briefed them as part of the commanding general's update brief. This was the third briefing in the daily commanding general's update that was broadcast by video teleconference (VTC) to all other parties involved in the operation. The safety briefing was preceded by the operations update and the force protection briefing, which was a very natural flow. The safety threat assessment brief included an assessment of threats facing the force within the next 24-36 hours and included risk management information as well as standards to help reduce the risks. Brigadier General B.B. Bell, the USAREUR (Forward) Chief of Staff, provided a vision of information management that built a full court press to get the word out on hazards and the measures used to control them. The information was targeted at subordinate battalion size units to include the Task Force in Bosnia-Herzegovina. Recommended control measures were implemented and supervised. The commanding general's daily update VTC was the primary mode for information distribution, followed by e-mailing the threat assessment briefing to all units, and followed by delivering hard copies to all units moving through the process who had no access to e-mail. Safety was also part of a STAR briefing that traveled to each unit in Bosnia-Herzegovina prior to their redeployment and provided the unit with information on how to redeploy and the hazards associated with the various stages of the process.

The redeployment process was the opposite of the deployment. Units came from Bosnia-Herzegovina through the ISB in Slavonski-Brod, Croatia then proceeded through the ISB in Hungary and back to Germany. Units redeploying went through a seven-day process at the ISB in Taszar, Hungary with a new unit on a different part of the process each day. Built into the process was down time to allow the soldiers to unwind and prepare to rejoin their families and friends. The marshaling area was a portion of a runway blocked off for vehicle parking for several hundred vehicles. It became so congested that roads, taxiways, and additional ramp space were used for vehicle parking. This became one of the biggest hazards with hundreds of vehicles and people moving during the hours of darkness. Emphasis was placed on the use of reflective items and chemical lights to highlight soldiers in the dark. A greater concern for the whole operation was convoy traffic throughout Hungary, Croatia, and Bosnia-Herzegovina. Even though the roads were improved, the large vehicles were still moving long distances during the hours of darkness. In this phase, vehicle drivers also faced an increased number of civilian vehicles on the roads than were present during the deployment in 1995.

A special assessment team was put together with Roy Valiant, Helma German, Carter Boggess and Major Mobley (ASAD). This team surveyed all processes involved in the redeployment at the ISB in Taszar, Kaposvar and Kaposujlak, Hungary. The surveys were an upstream evaluation of the processes before an accident occurred to support the next day or week's operations. Naturally, any hazards present for the unit going through the process were brought to the attention of the using unit, 7th Combat Support Group, and USAREUR Forward, Chief of Staff. Corrective measures were taken to resolve the hazards and correct the system to remove the hazard for the future customers. This team ran 24 hour-a-day operations to evaluate the process at all hours with all types of units on a continuing basis for the six weeks of the heaviest redeployment activity. It is widely believed that the efforts of this team contributed significantly to the safe redeployment. It would not have been successful if the USAREUR Forward and 7th Combat Support Group staffs had not reduced and controlled the hazards and systemic defects found by the team. The areas covered by the team included ammunition down load site, four rail heads, marshaling area, vehicle parking and maintenance areas, two

185

airfields, ranges, and a life support area of 500 tents each using two kerosene heaters, and the wash rack (used in freezing weather).

The Force Protection Working Group that was so needed through the first two phases of this operation was now more important than ever. Safety was an integral part of the working group. The USAREUR (Forward) G2 (Intelligence), Colonel Maxie McFarland, oversaw force protection and within the G2 a force protection office was established. This office was run by Lieutenant Colonel Geoff Irons, the deputy Corps G2. The force protection working group came under the force protection office and dealt with issues including health, hazards, security, terrorism, etc. The USAREUR (Forward) G2 carried items from the force protection working group directly to the USAREUR (Forward) Commander. Two major elements are noteworthy. First were the Force Protection Surveys. Specific teams from the Force Protection Office conducted these surveys. They were instrumental in identifying hazards, safety and otherwise, present in all organizations and locations. Steps were then taken to reduce the hazards or eliminate them completely. Secondly, a convoy assessment team was put together from this group that focused its attention on the routes, convoy procedures, rest stops, and actual conduct of convoys. This group spent many long days on the road identifying hazards and working to reduce the risks for military convoys.

Summary

The safety program developed and used during Operation Joint Endeavor is a living model and standard of a tactical safety program. The ASADs were an integral part of this program. Their presence seemed to solidify what they were all about, combat support for unit safety staff. After this experience, the safety professionals knew whom to turn to for assistance. The new model of the V Corps Safety Office allowed them to maintain personnel in three different locations pulling together for one program. This model was solidified upon their return to Germany. Each civilian safety professional made a commitment to the success of Operation Joint Endeavor as well as set the standard for safety during Operation Joint Guard, the follow-on mission. This commitment often meant 18-20 hour days seven days a week. Thanks to such dedication, many officers and NCOs alike will remember the service and support provided by these professional women and men. Some of these officers and NCOs will go on to positions as commander and Command Sergeants Major who will expect the maximum standard in the future from civilian employees, especially safety professionals, not the minimum. In addition, soldiers will better understand their role in accident prevention and how they fit into the whole process of preventing accidents.

Note: This article was originally published in the Center for Army Lessons Learned - News from the Front." July-August 2000 edition

Closing Thoughts

Thank you very much for taking the time to read these essays. I hope they will be of great use to you. I also hope you take the opportunity to use them in your life and work whenever possible. It is not cost effective to reinvent solutions to safety hazards each time they are needed when the work has been done and available for use. I also encourage Safety, Health, and Environmental (SHE) professionals to write and publish their own work so that together we can increase the body of knowledge for the SHE profession. Everyone's ideas are important and should be shared for the good of the profession.

In this book, I have included essays on training, hazards, the profession, and the professional. I have found them all applicable and I think these topic areas are important to every professional. This is a wide career field broken into specific program areas. I have only touched on a few here; however, I think these are some of the most important.

If you would like to help other readers out, please leave a review of this book on Amazon.com. Your rating and review will help them decide to read or not read this book.

From my other books, I recommend Basic Construction Safety and Health. You can see it at the following URL Basic Construction Safety and Health: Fanning, Mr. Fred: 9781492982210: Amazon.com: Books

About the Author

After a successful career as a Federal Employee that included over twenty years in safety and occupational health. I started writing part-time. My published work includes the peer-reviewed book Basic Safety Administration: A Handbook for the New Safety Specialist in its second edition. I also authored two editions of the peer-reviewed chapter, Safety Training and Documentation Principles published in the bestselling, Safety Professional Handbook, and the Safety Professional Handbook Management Applications, both edited by Joel Haight, Ph.D., CSP. I co-authored the peer-reviewed chapter Safety Training with Christine Fiori, Ph.D., PE, published in the bestselling Construction Safety Management and Engineering, second edition edited by Darryl C. Hill, Ph.D., CSP. The American Society of Safety Professionals Traditionally published my book and chapters.

I self-published another eleven books using Kindle Direct Publishing. Seven of these books are available in paperback and Kindle formats. Four of those books are available only in Kindle format. I have authored over fifty articles in various publications on safety and occupational health and project management. I have earned several writing awards for my non-fiction work and one for my fiction work. I have self-published two novels, A Walk Among the Dead and my most recent Mystery at Devil's Elbow.

I am an Emeritus Professional Member of the American Society of Safety Professionals. I was selected as the Safety Professional of the Year for the Northern Virginia Chapter of this Society. I am also a member of the Non-Fiction Writers Association. I held the Certified Safety Professional (CSP) designation for ten years. I also earned master's degrees from National-Louis University and Webster University.